CHICAGO MAPMAKERS

Chicago Mapmakers

Essays on the Rise of the
City's Map Trade

Edited by
Michael P. Conzen

The Chicago Historical Society
for the
Chicago Map Society
1984

© 1984 by the Chicago Historical Society

Library of Congress Cataloging in Publication Data
    Chicago mapmakers.
    Includes bibliographies.
    1. Cartography—Illinois—Chicago—History.
I. Conzen, Michael P. II. Chicago Map Society.
GA421.C48 1984   338.7'61526'0977311   84-5853
ISBN 0-916789-01-2

Preface

Since its formation in 1976, the Chicago Map Society has encouraged study of American mapping in the context of the general history of cartography. In 1978, Patricia A. Moore, a member, convinced the Society to commission a series of lectures on Chicago mapmakers, based on original research, in order to cast new light on the origins and early development of Chicago as a great center of mapmaking. That initiative resulted in several addresses to the Society and further research on the subject, selections from which have been brought together here with the aim of reaching a wider audience.

It has been common in Europe to focus scholarly attention on the map trades of the region's major cities in elucidating certain aspects of the history of cartography. In North America such an approach has been little developed, and consequently we lack a clear understanding of the role particular cities have played in incubating and advancing cartographic work in America over the last two centuries or so. While Philadelphia, New York, Boston, and to a lesser extent Baltimore emerged as predictably early centers of American mapmaking, considering their colonial origins, size, and well-established commercial connections, Chicago succeeded with astonishing speed to earn the title of cartographic capital of the United States by the end of the nineteenth century. This invests the Chicago map trade with more than usual interest, and this collection of essays is offered, then, as a contribution to the history of American commercial mapping.

The appearance of this volume owes a good deal to the support and assistance of several individuals and groups besides the authors. David Woodward and David Buisseret, former and present directors respectively of the Hermon Dunlap Smith Center for the History of Cartography at the Newberry Library in Chicago, gave valuable encouragement to the project at critical stages. Timothy Jacobson, director of publications at the Chicago Historical Society, has played an indispensable role in arranging publication of the volume, contributing editorial expertise, and making available the considerable illustrative resources of the Society, including its outstanding collection of Chicago cartography. The keen support of the board of directors and members of the Chicago Map Society as sponsors of this publication is gladly acknowledged.

All the authors appreciate the extensive assistance they have received from the staffs of the Chicago Historical Society, particularly Grant Dean, map curator, and the Newberry Library, especially the Department of Special Collections. R. G. Dun and Co. are gratefully acknowledged for permission to use manuscript credit records on selected Chicago mapmakers in the collections of the Baker Library, Harvard Business School. Cynthia Peters wishes to thank Rand, McNally and Company for permission to examine the firm's archives and acknowledges the special assistance of staff members Joe Rocky and Paul Tiddens. She also thanks Howard Winger, of the University of Chicago, and David Woodward, now of the University of Wisconsin, for their initial guidance. Michael Conzen thanks Richard Stephenson, head of the Reference Department of the Geography and Map Division, Library of Congress, for many useful discussions and scholarly courtesies, and also Kathleen Zar, map librarian at the Regenstein Library of the University of Chicago, for varied and valuable assistance with this undertaking.

MPC

Contents

4	Evolution of the Chicago Map Trade: An Introduction MICHAEL P. CONZEN
12	Chicago's First Maps GERALD A. DANZER
23	Rufus Blanchard: Early Chicago Map Publisher MARSHA L. SELMER
32	George F. Cram and the American Perception of Space GERALD A. DANZER
47	Maps for the Masses: Alfred T. Andreas and the Midwestern County Atlas Map Trade MICHAEL P. CONZEN
64	Rand, McNally in the Nineteenth Century: Reaching for a National Market CYNTHIA H. PETERS
73	Made in Chicago: Maps and Atlases Printed in Chicago Before the Fire ROBERT W. KARROW, JR.

ILLUSTRATIONS

5, CHS, ICHi-18054, from *Edward's Annual Director to the Inhabitants...in the City of Chicago* (1869); 8, 9, courtesy of Michael P. Conzen; 11, CHS, ICHi-18168, from *John C.W. Bailey's Chicago City Directory* (1866-67); 13, CHS, ICHi-15570; 14-15, CHS, ICHi-18061, from *A Business Advertiser and General Directory of the City for the Year 1845-46* (1846); 17, CHS, ICHi-18072; 18, CHS, ICHi-18069; 19, CHS, ICHi-18071; 21, CHS, ICHi-18070; 24, CHS, ICHi-18060, from *Case & Co.'s Chicago City Directory* (1856-57); 25, CHS, ICHi-18074; 26-27, CHS, ICHi-18075; 28, CHS, ICHi-18073; 29, CHS, ICHi-18029; 30, courtesy of The Newberry Library, Chicago; 33, from *Cram's Unrivaled Atlas of the World* (1889), courtesy of the University of Illinois at Chicago, The Library, Map Section; 34, CHS, ICHi-18065, from *Cram's Superior Reference Atlas of Illinois and the World* (1906); 37, from *Cram's Unrivaled Atlas of the World* (1889), courtesy of the University of Illinois at Chicago, The Library, Map Section; 39, CHS, ICHi-18031; 40, from *The Columbian World's Fair Atlas* (1893), courtesy of Gerald A. Danzer; 43, CHS, ICHi-18067, from *Cram's Superior Reference Atlas of Illinois and the World* (1906); 44-45, CHS, ICHi-18157, from *Cram's Superior Reference Atlas of Illinois and the World* (1906); 46, CHS, ICHi-18084, from *An Illustrated Historical Atlas of the State of Minnesota* (1874); 49, CHS, ICHi-09410; 51, CHS, ICHi-18164, from *Atlas of Tazewell County, Illinois* (1879); 52, from *How 'Tis Done: A Thorough Ventilation of the Numerous Schemes Conducted by Wandering Canvassers Together with the Various Advertising Dodges for the Swindling of the Public*, courtesy of The Newberry Library, Chicago; 55, CHS, ICHi-18079, from *An Historical Atlas of the State of Minnesota* (1874); 56 above, CHS, ICHi-18078, from *Atlas Map of Fulton County, Illinois* (1871); 56 below, CHS, ICHi-18081, from *Atlas Map of Fulton County, Illinois* (1871); 59, CHS, ICHi-18082, from *Atlas Map of Fulton County, Illinois* (1871); 60, CHS, ICHi-18162, from *Atlas Map of Fulton County, Illinois* (1871); 62, CHS, ICHi-18087, from *Atlas Map of Fulton County, Illinois* (1871); 65, CHS, ICHi-18058, from *The Rand-McNally Official Railway Guide and Hand Book* (1880); 67, courtesy of the Everett D. Graff Collection, The Newberry Library, Chicago; 68, CHS, ICHi-18057, from *The World of Rand McNally* (1956); 69, CHS, ICHi-18056, from *Rand, McNally & Co.'s Handbook of the World's Columbian Exposition* (1893); 70, CHS, ICHi-18160, from *The Rand-McNally Official Railway Guide and Hand Book* (1880); 71, courtesy of Michael P. Conzen.

Evolution of the Chicago Map Trade: An Introduction

By Michael P. Conzen

THE EXPLORATION, settlement, and expansion of America would have been impossible without maps. So vast and varied a territory needed reducing to a level of human comprehension that the map alone, through its ordering of natural phenomena in their proper spatial relationship, could provide. Even as European colonization spread a new kind of civilization across the continent, new maps of all kinds and at all scales were needed to keep people abreast of developments and offer sound geographical information for future planning. The men—and occasionally women—who made these maps, the methods they used, and the appearance they gave them, are vital aspects of the role maps have played in national development and cultural definition.

Maps traditionally have been far more difficult to produce in printed form than book text or even some types of drawing and painting. Book text requires an author, typesetter, printer, and bookbinder. But with cartography, particularly in the era before photolithography, a typical map required a surveyor or compiler, a draftsman, an engraver, a printer, a map colorer, and map mounter or finisher. Almost any of these contributors to the map's production might have acted as its publisher too. Not surprisingly, bibliographers over the years have had great difficulty identifying all the contributors to specific maps, and even to whom they should assign primary authorship. More than with most types of publishing, mapmaking thus has long been a communal enterprise, often split between compilation and surveying "in the field" and specialized craft skills and machine production at the printing stage. It also has concentrated generally in urban centers of some size.

Since the first days of the printing press and the woodcut, large cities have been crucial to mapmaking because as centers of technical invention and business innovation they have harbored the largest concentrations of artisans and machinery specializing in map production. Also, it is only in such places that rural as well as urban demand for maps would have accumulated sufficiently to support them. In the United States, indigenous map production arose to compete with the output of London and other European map centers by the end of the eighteenth century. It concentrated naturally in the established metropolises of Boston, New York, Philadelphia, and later Baltimore. By the middle of the nineteenth century, New York clearly dominated in sheer map output, but second-place Philadelphia was developing some map specializations, such as county landownership mapping, for which it was the prime center. Developments elsewhere reflected the shorter history of settlement in the interior while Cincinnati, St. Louis, and then Chicago grew and matured enough to offer the specialized facilities needed for local mapmaking.

There has been little scholarly inquiry into the emergence of map trades in particular American cities, and accordingly, a general picture of their process of growth and entry into competition for the national map market is difficult to reconstruct. One fruitful approach is to examine the history of individuals and firms who played significant parts in the building up of a city's map business. This special collection of essays traces the outline of this development in Chicago through the mapping careers of Rufus Blanchard, George F. Cram, Alfred T. Andreas, and the firm of Rand, McNally and Company. Collectively, they span virtually the entire period in the nineteenth century when maps bearing a Chicago imprint were published, and they illustrate much of the variety within Chicago mapmaking, both in their approaches to the business and in the types of maps they produced. This introduction to the evolution of the Chicago map trade provides a broader setting for their activities.

Michael P. Conzen is associate professor of geography at the University of Chicago.

Richard Edwards, publisher of Edward's Chicago Directory, *seated before a typical wall map illustrating the convergence of railroad routes on Chicago. Such maps were an essential part of office furnishings in nineteenth-century Chicago.*

Chicago's Map Trade

The Chicago Map Trade: The Incidental Phase

Chicago in its infancy lacked almost all technically sophisticated trades and services, and libraries and bookstores were absent or quite rudimentary during the city's first decade. Chicago's first business directory in 1839 listed one bookstore under the proprietorship of S. F. Gale, a wood and metal engraver named Shubael Davis Childs, and a bookbinder, Hugh Ross. Occasionally during the 1830s and 1840s, maps were prepared of the city plat, Fort Dearborn, and other local sites, but they remained in manuscript form or were printed in Washington, D.C., or New York for government reports or real estate interests. Maps used by early Chicagoans were either the few manuscript plats on file in the local land office and courthouse and a few circulating hand-drawn copies, or printed maps of western regions imported from the East.

Actual map printing and publishing in Chicago before mid-century was highly intermittent. Juliette Kinzie is probably responsible for the first printed map, a primitive effort titled *Chicago in 1812* which appeared in her *Narrative of the Massacre at Chicago*. The next certain published map from Chicago came in 1847, a *Map of the Route of the Galena and Chicago Union Railroad* prepared by Richard P. Morgan for his report on the survey for the route and engraved by Childs's erstwhile associate, Roswell N. White. An early general reference map of *Chicago and Vicinity*, while compiled by James H. Rees and Edward A. Rucker, Chicago land agents, and drafted by local surveyor William Clogher, was printed in 1849 by the St. Louis lithographer Julius Hutawa. By this time, Chicago boasted five bookstores and stationers, a bookbinder, two engravers (including one who worked with copperplate), and about eight printers, and yet most maps of Chicago or general maps used there were still produced elsewhere.

The Take-off Phase

The dominance of eastern maps in the Chicago trade, especially those from New York and Philadelphia, continued well past the Civil War. It reflected their high quality of information and execution, their well-established distribution channels, and the paucity of map printing expertise in Chicago. But the Lake Michigan town burgeoned in the early 1850s; its hinterland was vast, and the tides of migration swept in new demand for maps and the artisans to make them. Three different maps of the area involving local cartographers appeared in 1851, and by 1853 the arrival of Henry Acheson and Edward Mendel, two lithographers who specially advertised their mapmaking skills in the city directory, served to anchor the business firmly in the booming metropolis. The lithographic process was suited ideally to printing graphic material such as maps, and its cheapness, when compared with copperplate methods, made it especially attractive in a raw city near the frontier. From then on, maps were routinely manufactured as well as compiled and sold in Chicago itself. Since booksellers had been stocking eastern maps for resale since the early days, it is natural that some book dealers acted as local map publishers when opportunity arose. David B. Cooke, the Burleys, and Keen & Lee published maps in the mid-1850s, often in collaboration with Mendel or one of the other Chicago lithographers. When Rufus Blanchard arrived in 1854, he set up his Chicago Map Store very much in the style of a general bookstore and eastern map retailing operation, though he quickly entered the business of direct mapmaking and publishing.

Chicago mapmaking thus began in earnest in the early and mid-1850s. Acheson and, more particularly, Mendel generated a rapidly increasing list of map publications, and between 1851 and the Civil War 80 percent of the maps made of Chicago were local imprints. While New York had dominated the supply of maps of Illinois before 1854, Chicago thereafter captured one-third of that market before the war. Clearly, as a new mapping center seeking to enter national competition, Chicago made most headway with local and regional coverage. By 1855 Chicago's city directory sported a special "Maps and Charts" section in its classified business listings, and the following year it carried the first advertisement for an individual map—E. Hepple Hall's *New Map of Chicago*—available at the directory office for 50 cents. Local publishers also began to diversify the kinds of maps they produced. The plan for Chicago's first sewerage system emerged in 1855 from Mendel's office, while Acheson printed the Illinois Central Railroad's map of its 2.5 million acres for sale. Later in the decade, other railroad maps appeared, showing route networks; a post office chart for Illinois and a map to accompany a report on the Illinois coalfield also were published. And by 1859 Chicago was ready to challenge New York and Philadelphia in maps of the whole country with *D. B. Cooke and Co.'s Railway Map of the United States and Canada*.

The Diversified Phase
The decade from 1861 to 1871 saw the consolidation of Chicago's map trade. Map publishers increased from four to six and specialized services like map coloring and map finishing came to be separately advertised. The establishment of George W. Terry's map mounting business around 1865 underscores the expanding demand for maps mounted on cloth, especially those on rollers as wall maps for office and classroom. More notable, however, was the growing diversity of map production itself. Both the first commercially published geological and mineral map of Illinois, and an important early regional map produced in Chicago, a *Map Showing the Position of Chicago, in Connection with the North West...*, were issued in 1861 from Mendel's lithographic office. In the same year the first comprehensive political map of Illinois, by Moses H. Thompson of Geneva, showing congressional and legislative districts, was printed by Charles Shober, another lithographer who became a prolific Chicago mapmaker.

Real estate maps of Chicago had been a fixture since the city's inception, but early ones emanated from New York, the source of so much speculative interest. Land agent James Van Vechten in 1863 initiated a long-running series of wall-map-sized reference plans of Chicago, and two years later Frederick Cook, a civil engineer, produced the city's first fire insurance atlas, featuring detailed, hand-colored maps with a scale of one inch representing forty feet. Edward Mendel, who printed Cook's atlas, also developed a large rural business during the 1860s lithographing county landownership maps for surveyors around Illinois and the Midwest. By 1867 he turned out an exquisite county atlas of rural real estate plats for Lyman G. Bennett's *Map of Winona County, Minnesota*, with delicate hachures accenting the county's rolling terrain and Mississippi bluffs. Charles Shober also profited from the county map and atlas trade in the hinterland, as did the newly established firm of Warner and Higgins (later Warner and Beers) that came to Chicago in 1869.

A year earlier, Blanchard published his *Cabinet Map of the United States and Territories*, the first national map completely produced in Chicago—lithographed by Shober—while George F. Cram, Blanchard's nephew and brief apprentice, began his own cartographic publishing career. By 1870 the Western Bank Note and Engraving Company issued an election map of the city, showing ward boundaries and voting totals for each ward in the 1870 election. The decade had been a crucial one, then, for Chicago mapmakers, both in aggressively developing varied and voluminous local and regional work as well as making opening bids for a more national market. But still the sales of Chicago-made maps were largely within the Midwest, and eastern maps and atlases, particularly those for educational and reference use, held sway throughout the country, as well as in Chicago.

The Specialized Phase
These patterns of competition and business organization might have continued longer and changed more subtly had it not been for the Chicago Fire in October 1871. The fire played havoc with the city's printing establishment. Overwhelmingly concentrated in the central business core around Lake, Clark, and Randolph streets up to that time, most printers, engravers, and publishers sustained heavy losses in the disaster. Several businesses simply folded, but a surprising number regrouped, absorbed the shock, and survived—without a doubt a consequence of general business confidence in the city and lenders' willingness to extend unprecedented credit. Characteristic was the experience of the Lakeside Publishing and Printing Company. Formed as the successor to Church, Goodman and Company in November 1870 with a capital of $500,000, the firm sunk $115,000 into the early stages of a grandiose building. Construction had reached to the fifth story when the fire struck, wiping out all but $17,000 worth of building foundations. Undaunted, the firm recommenced building and took occupancy in November 1872. This new Lakeside Building could accommodate many firms.

The Lakeside Building's role in relocating under one roof several firms involved in mapmaking symbolizes a broader reorganization of the

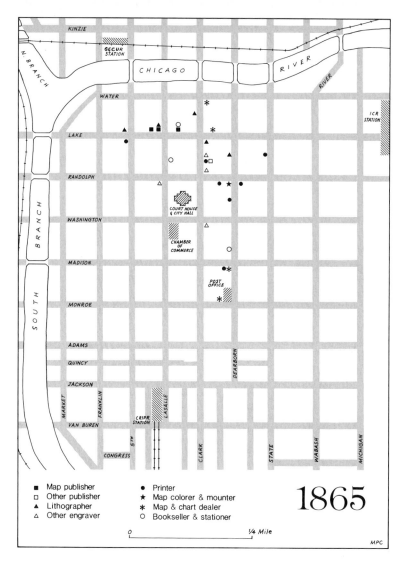

Changing location of Chicago mapmaking business. Before the Fire of 1871, the specialized services required for map publishing were overwhelmingly concentrated in the central business core around Lake, Clark, and Randolph streets.

Chicago map trade hastened in large part by the fire. Besides housing the printing company that gave it a name, the Lakeside Building became the business address for the atlas publishers Warner and Beers and an off-shoot company of theirs, Higgins, Belden and Co. In addition, there was a new atlas man in town, Alfred T. Andreas, whose earlier business in Davenport, Iowa, while highly successful, was not successful enough and needed the contacts of a large city. Charles Shober moved into the Lakeside Building, too, as did August Maas, the wood engraver, A. J. Cox and Company, bookbinders, and A. H. Reeve with his gold beating works. While major map enterprises like Blanchard's and Mendel's relocated to more spacious quarters in a redeveloping business district, the impetus towards industrial integration, and especially towards atlas making, provided by this specially outfitted building was decisive.

In the next few years, county, and later state, atlases poured forth from the Lakeside Building and must have encouraged George Cram and Rand, McNally and Company—then entering mapmaking through railroad guide route maps—to realize that Chicago represented a sound location for the national atlas market. While the county and state atlases helped make Chicago the premier subscription publishing center in the country, outstripping Philadelphia in cartographic output and New York in the pure book trade, Cram and Rand, McNally and Company made Chicago the center of atlas trade sales (over the counter and by contract). The diversity of Chicago map publishing continued after the fire, but with new emphasis on high-volume, general atlas production for business, home, and school use. Employing cerographic printing methods brought to Chicago originally by Blanchard, although lightly used by him until a later period, Cram and Rand, McNally turned out millions of reference atlases in a range of prices that radically loosened the New York grip

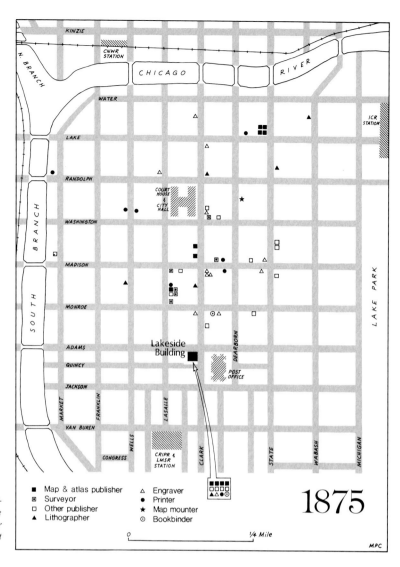

The Fire played havoc with the city's printing establishment. As firms recouped, some took the opportunity to reorganize under one roof in the new Lakeside Building at Clark and Adams streets.

on this market. Cram's first atlas appeared in 1875, and Rand, McNally's first *Business Atlas* in 1876, and from there neither firm looked back.

The atlas impulse in Chicago mapmaking reached even further. Chicago after the fire made itself "conspicuous, and almost ridiculous" by its rapid physical growth, as Bessie Louise Pierce has quoted one visitor as saying. The necessity of keeping up with the city's extraordinary building boom spawned a series of major real estate atlases and fire insurance atlases of the metropolis. Otto Pelzer, a one-time Blanchard associate, produced the first in 1872, G. R. Hoffman produced a sequel in 1875, and H. R. Pasge and Company a third in 1879. Meanwhile, Charles Rascher entered the property atlas field with his 1877 *Fire Insurance Map of Chicago*. Unlike all other kinds of atlases, fire insurance atlases usually introduced the "correction service," which for a fee supplied annual paste-on kits to update the map information, and thus bound publisher and customer in a long-term relationship. Similar developments occurred in all other major cities, and the Sanborn and Bromley companies were able later to create multi-city atlas services. But in Chicago in the 1870s and 1880s, it was local men who capitalized a major new sphere of mapmaking, born of and catering exclusively to metropolitan gigantism.

The decade after the fire saw a steady stream of city and metropolitan guide maps, railroad maps, and maps showing off the park and boulevard system of Chicago, in addition to now standard political, reference, and pocket maps of states and regions. It was a mobile era, the railroads were in their prime, and everyone needed to know where they were going. What stands out about this decade, however, is how the Chicago map trade specialized in several key types of mapping, especially atlases. As this specialization took root, it had consequences for the size of the industry and its structure.

Chicago's Map Trade

The Corporate Phase

Between 1880 and the end of the century the number of map and atlas publishers in Chicago actually decreased from eleven to ten, but output soared. Economies of scale in production methods allowed large and successful companies not to extinguish competition but rather to limit the potential scope for newcomers. Hence the new names in mapmaking by 1900 were either national firms with muscle and a Chicago office like the Sanborn-Perris Map Company, or large local publishers like Reuben H. Donnelly, a successor to the Lakeside Publishing and Printing concern, who had cornered the Chicago real estate atlas market.

Insurance and real estate atlases continued along with geographical reference atlases to dominate the map trade in the closing decades of the century. Most large firms were well established, and those that adapted to new technology, such as Jacob Manz and Company and August Zeese and Company in the case of electrotyping, prospered in their chosen specialties. George A. Ogle emerged in the 1880s as a spirited successor to Alfred Andreas in county atlas publishing, which experienced a rebirth in the 1890s. And William Wangersheim, under his own name and as the proprietor of the later Standard Map Company, managed to carve a large niche in the utility map publishing sphere of the trade. By doing much subcontract work for Ogle and others, he supplied some healthy competition in general map printing to Rand, McNally and Donnelly. Such utility printing by the close of the century had come to include maps of Chicago's complex transit system, downtown locator maps, suburban maps—stimulated often by the city's political annexations—and maps occasioned by special events such as the 1893 world's fair.

Mapmaking in Chicago after about 1880 is striking less for the types of maps published than for the manner in which the principal firms were becoming major corporations. Long gone was the pivotal initiative of Rufus Blanchard in his early career, who never let the creative cartographic side of the business out of his personal control. Gone also was the flamboyance—and the pecuniary instability—of an Alfred Andreas. In his heyday, 1875, less than one-third of the mapmaking firms were limited companies, whereas twenty years later three-quarters of them were. Mapping by 1900 was indeed big business, and its practitioners were organized accordingly.

The Special Character of Chicago Mapmaking

Chicago mapmaking has left an indelible mark on the American self-consciousness, if only because generations of Americans have seen their nation through the cartographic images shaped in the traditions of Chicago-made maps and atlases. By European and even East Coast comparison, Chicago cartography did not develop on the basis of aesthetic design or flawless execution. This was a matter of timing and temperament. Chicago was settled too late to participate in the high era of elegant copperplate mapmaking, and the restlessness of its business life and the vastness of its hinterland was far more suited to the speed and cheapness of lithography and wax engraving. Indeed, Chicago's ascendancy in mapmaking owes not a little to this latter printing technology, which it made its own.

What nineteenth-century Chicago maps lacked in fanciful appeal, however, they made up in their sheer abundance, their currency, and their appeal to the pocketbook. To the extent that Chicago mapmakers proved so aggressive in steering the city toward cartographic preeminence in this century, they created an industry and a published record that was technically and organizationally at the cutting edge of American map development. The application of photography to map printing in the nineteenth century, as the historian of cartography Arthur Robinson has noted, allowed newcomers to the art of cartography to play with map design. Design conventions developed over centuries became disposable, and simple functionalism—often dictated by the limitations of machinery or time—triumphed. This was not a moral failure, but it offers a keen insight into the tension between utility and creativity.

Chicago, more than any other city, defined the developing scope of American commercial mapmaking in the second half of the nineteenth century. Until a more extended inquiry is made into the nature of the Chicago map trade, these essays on the accomplishments and experience of several important Chicago mapmakers may serve to illustrate the bases of its long-term success and the prospects it faced at the dawn of a new century.

Charles Shober was one of the most prominent figures in Chicago mapmaking. This page from John C. W. Bailey's Chicago City Directory, 1866-67, *advertised Shober's services as a lithographer.*

Chicago's First Maps

By Gerald A. Danzer

Between the lakes and the great western river system lay the site that would become Chicago. Early maps foretold its advantages.

THE PLACE encouraged the making of maps from the very beginning. At the inception of European contact, explorers Joliet, Marquette, and La Salle all made maps to record their journeys beyond the lakes. The first two were particularly struck by the advantages of the place that their Indian escorts called Checagou. In the fall of 1673, after traversing the Chicago portage between the great river system of the West and the "Lake of the Illinois," each adventurer commented on the advantages of the site. To make their point, both soon composed maps that highlighted the merits of the place where the water routes of the two physiographic provinces reached out to touch one another. The controversies over what happened to these early French maps need not detain us. The point is simply this: the geographical setting of Chicago immediately called for maps to illustrate its advantages. By the end of the seventeenth century Checagou appeared with some regularity on the published maps of the North American interior. Thus the place named Checagou antedates the settlement by at least a century in the cartographic lexicon.

In the century following 1673, Chicago served as the location for various occasional endeavors: a Jesuit mission, a military campsite, an Indian village, a council grounds, and a rendezvous for traders in the region. Whether any of these different uses called for crude maps of site or direction, we do not know. Perhaps a simple reference to the point at the southwestern extremity of the Great Lakes, where their waters met both the prairie and a major branch of the Mississippi River, was enough of a description. In a word, this place was Chicago.

Gerald A. Danzer is associate professor of history at the University of Illinois at Chicago.

The topography of the site was so flat that the boundaries between land and water shifted with the seasons. The slight local relief was provided by occasional groves of oak and hickory, which broke up the monotony of the tall grass prairie. It was not until the region had changed hands several times that a surviving map, plat, or view of Chicago appeared. The Treaty of Paris in 1763 transferred the territory from the French to the British Empire. Two decades later another Treaty of Paris passed the land on to the new United States of America. Then in 1795, Article III of the Treaty of Greenville canceled Indian claims to fifteen scattered sites in the Old Northwest, including "one piece of land six miles square, at the mouth of the Chicago River." The government's purpose in acquiring these scattered parcels of land was to build forts to help the young nation secure its claims to the region.

The construction of Fort Dearborn in 1803 led directly to the earliest surviving graphic representation of Chicago. Captain John Whistler, commander of the company of soldiers who both erected and garrisoned the early fort, sketched a plan of the site in 1808, the earliest entry in the catalog of maps made in or of Chicago. Whistler's pen-and-ink drawing combined an outline map of the site and a plan of the fort with drawings of the military buildings as well as several of the traders' cabins which lined both the river and the road to Detroit. Whistler's plan is apparently the sole surviving graphic record of the settlement around the first Fort Dearborn.

Captain John Whistler, it also might be noted, received his artistic and cartographic training in Great Britain. As a young soldier in Burgoyne's army, he was taken prisoner by the patriot forces after the Battle of Saratoga. Later he settled in America, married an American, and enlisted in the army of the new nation. He rose rapidly in the

Above: *The earliest surviving map published in Chicago (1844), showing Fort Dearborn and the surrounding settlement.*

Preceding Page: *The portage of Chicago, which explorers had documented, was featured on many early maps, such as this copperplate example published in France in 1755.*

ranks while fighting in the Indian wars that led to the Treaty of Greenville. Thus it is possible to link directly an early map and view of Chicago with a heritage of professional cartographic and artistic training in Europe.

After the destruction of the fort in 1812, the fame of the place grew. Several graphic representations have survived from the period between the rebuilding of the fort in 1815 and the founding of the town in 1830. By treaties made with the Indians in the area after 1815, the United States government acquired a strip of land between the Chicago and the Illinois rivers. This cession secured the route celebrated by the French explorers. Combined with the earlier acquisition for the fort, these lands were assigned to the United States Land Commissioners who started to survey the tract about 1818. An early manuscript map based on their notes as well as several plats of the government survey dated 1821 were presented as evidence in the celebrated court case of *George C. Bates v. the Illinois Central Railroad Company* in 1859.

IN 1812.

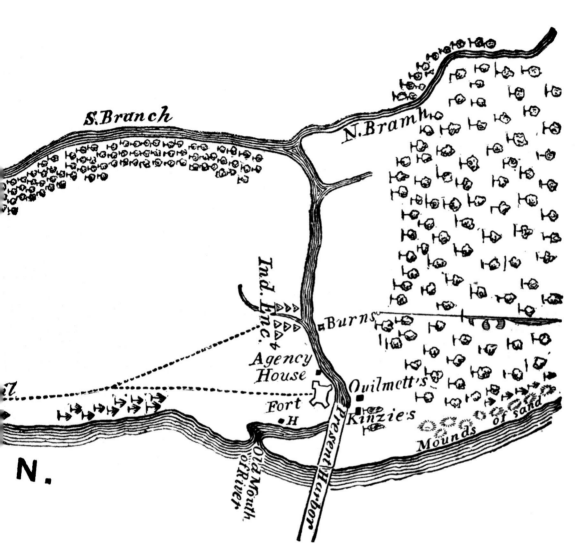

Military authorities, land office surveyors, and various engineers concerned with the development of the harbor at the mouth of the river continued to draw maps of Chicago for various agencies of the federal government after 1818. Major attention was focused on the shifting sandbar that interfered with navigation between the lake and the river. This neck of sand figured prominently in the sketch which Henry Rowe Schoolcraft made of Chicago in 1820. Schoolcraft's view, looking west from a point just offshore, with Fort Dearborn on the left and with the Kinzie house and trading post on the right, became the classic image of early Chicago. A print based on this sketch appeared much later in Schoolcraft's six-volume study of the Indian tribes in 1854.

The troublesome sands blocking the mouth of the harbor remained a focus of attention from 1818, when the soldiers at the fort made a shallow cut through the sandbar, until the late 1860s. The shifting contours at the mouth of the river can be traced in various engineers' reports to Washington. Several of these, printed in various government documents, are in the *Checklist of Printed Maps of the Middlewest to 1900*. The earliest printed accompanied the annual report of Thomas Jefferson Cram of the United States Engineering

Chicago's First Maps

Corps. Reprinted as a Senate document in 1839, the map originally was drawn by Captain James Allen in 1837.

Copies of the harbor maps, whether in printed or manuscript form, were necessary navigational aids. One, a nautical chart of "Chicago Harbor, 1843 from a survey by Lieut. W. H. Warner," has been preserved at the Chicago Historical Society. Sydney F. Durfee, the city's first harbor master, prepared a smaller "Sketch of the Chicago Harbor." It shows only the approach to the harbor, indicating various depths of the channel, and emphasizing the sandbars south and east of the North Pier. Sailing directions appeared on the bottom of the chart.

Because we know that Durfee came to Chicago in 1845 and because the chart shows the sandbar adjacent to the North Pier that was cleared in 1852, this map can be dated at about 1850. A statement supplied by Durfee's widow in 1894, attached to the Chicago Historical Society's copy of the map, states that the block from which it was printed was supposed to be in the "historical rooms of the Chicago Book Company." This scant evidence makes the Durfee chart a candidate for the earliest nonhistorical map of Chicago to be printed in the city. A purist might object, however, on the basis that no terra firma, except the pier system, is represented on the chart.

In February 1830, at the very beginning of the harbor improvements and before the inception of the town, U.S. Assistant Civil Engineer Frederick Harrison, Jr., sent to Washington a manuscript titled "Proposed Plan for Improving the Mouth of the Chicago River." His superior, George Graham, then submitted the map to the Senate with the recommendation that "an act be passed authorizing the President to lay off a town" on the public lands "at the mouth of Chicago Creek." Section 9, immediately to the west of this tract, he pointed out, had been granted to the State of Illinois to support the digging of a canal. The state's section, he concluded, would "derive much benefit" by the extension of the town lots across the section line. This boundary soon became the eastern rather than the western border of the original town, because the state canal commissioners already had platted a portion of Section 9 as the original town of Chicago.

James Thompson's plat became the fundamental document in the formation of Chicago as an urban place. Up to this time, the land ceded by the Indian tribes had been surveyed and divided into townships and sections. Fort Dearborn stood on a military reservation comprising fractional Section 10 in Chicago's township. By an act of Congress in 1827, alternate sections in a strip of land ten miles wide along the proposed canal route were granted to the state to support the construction of the canal. Section 16, immediately south of Section 9, similarly was given by the federal government to local authorities for the support of education. After 1830 these various sections, or parts of them, were added to the original plat to form the matrix of public streets and private lots on which the city eventually was erected. But Thompson's plan remained the center of the metropolis, setting much of the pattern, location, and, in many cases, even the names of the streets and subdivisions.

If Thompson's design had such a great impact on the physical form the city would take, it is curious to note that he was not a Chicagoan and that his original survey never has been printed in its original form. Thompson was a skillful surveyor and a prominent local politician from Randolph County, Illinois. He came to Kaskaskia from South Carolina in 1814 to teach school. Soon he married and settled down on a farm. As a respected member of the community, he was elected a county commissioner and captain of the local militia. Along the way he did numerous surveying jobs for a variety of governmental offices. To his advantage, one of the three original canal commissioners also resided in Randolph County. It was probably through this connection that Thompson received the job of surveying the original town of Chicago.

His work for the canal commissioners, however, was only one job among many. According to family tradition the authorities wanted to give him lots in Chicago to pay for his surveying services, but he refused, receiving instead a well-bred mare which the commission somehow had obtained. After laying out the town as a typical river city, Thompson returned to his home county and became a judge. He also served in the Black Hawk War and contributed plans and surveys

Opposite: *Hand-drawn manuscript map by army engineer W. H. Warner, showing lake soundings and the sandbar at the mouth of the harbor.*

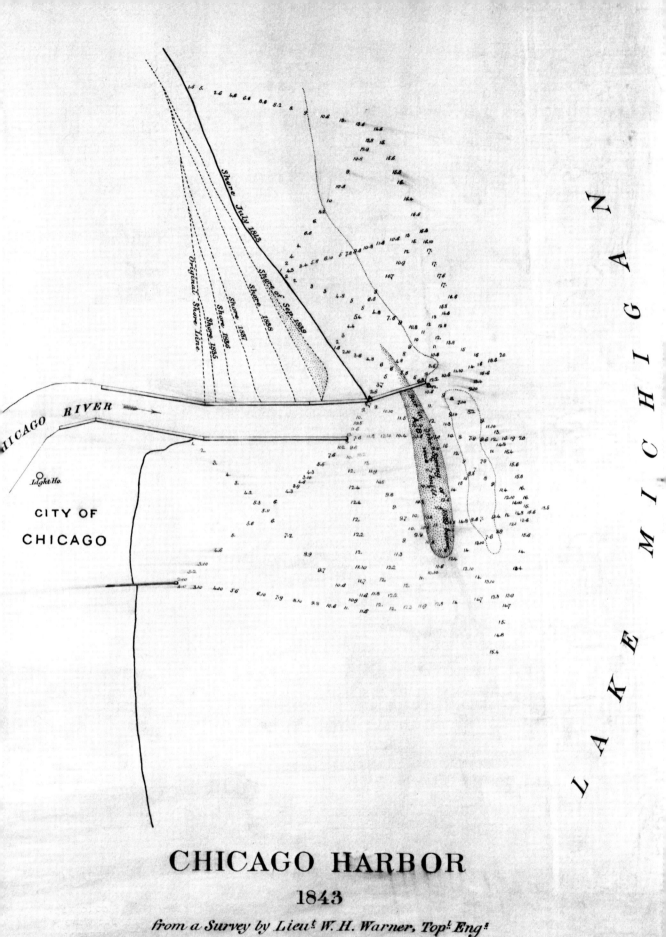

Chicago's First Maps

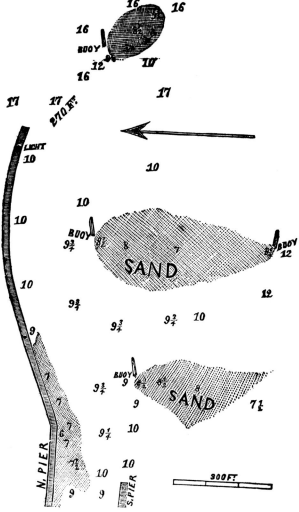

Harbor navigation map (c. 1850) prepared by Sidney F. Durfee, Chicago's first harbor master.

for several other Illinois communities. But he is remembered today largely because of the mark he set on the state's metropolis, which includes the presence of Randolph Street, named in honor of his home county.

In all probability, Thompson made several copies of his plan in 1830, at least one for the canal commissioners and one for public record. In the rush to sell some of the land, the plat apparently was not officially filed and recorded until 1837, by which time Chicago's land speculators already had published several maps of the city. The copy of record, of course, was destroyed in the Chicago Fire. The copy currently in the Chicago Historical Society's collections contains a certification by the "late commissioners" that this was "the identical map" used in the land auction of 1830.

As additional parcels of land were sold, subdivided, and added to the original town, the real estate interests of the city needed a map showing the whole community. Five such parcels had been added to Thompson's initial subdivision by 1834, including the large school section which had doubled the physical size of the community. Another factor also created a pressing demand for a town map: the legal community as defined by the recorded surveys was not apparent to an observer on the site. In 1834 the actual view was of a marshy prairie with roads, paths, and buildings distributed in random pattern according to the dictates of the early inhabitants. The settlement one saw in 1834 thus had very little in common with the town plotted by the surveyors.

Gradually, as streets were laid out and new construction followed the dictates of the surveyors, and some old buildings were moved onto their appropriate lots, the town became a closer match to its legal description. Once this was accomplished, largely by the late 1830s, maps describing how the town should look were no longer necessary. During the transition phase, however, maps were an absolute necessity. This helps to explain the sudden outburst of Chicago maps published during the two-year period between 1834 and 1836. Six real estate maps were printed in this brief span of time, followed by a lapse of more than a decade before the next published maps of the city appeared.

Maps of Chicago in the early 1830s were also in demand in eastern cities where investors were interested in purchasing western lands. Even before the first map of Chicago was published, various plats and plans probably circulated in manuscript form between those on the scene in Chicago and their financial backers in the East. The manuscript version of John Stephen Wright's 1843 map was sent to his uncle in New York so that the latter could follow the real estate transactions being made on his behalf. A fine copy of this early map, on tracing paper, is presently located in the National Archives and is readily

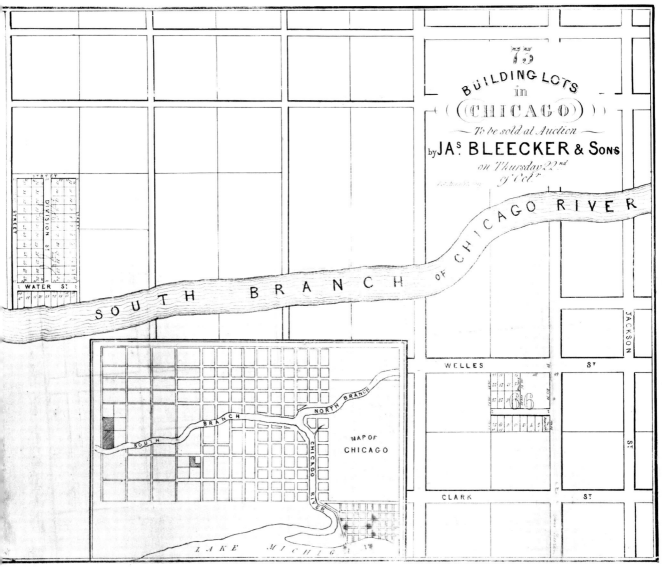

Real estate maps of Chicago were in demand in eastern cities among land speculators. Peter A. Mesier, a New York lithographer, produced several of these. His 1835 plat featured an inset map of the city showing the area of the lots for sale.

available in a modern reproduction. A map very similar to Wright's, but dated a year earlier, is located in the Chicago Historical Society. It carries a notation stating it is a copy of the map "sent to Washington to secure an appropriation to improve the harbor, March 2, 1833."

Wright's map, with some changes, soon appeared in New York City as a lithograph by the firm of Peter A. Mesier. It was not, however, the first published map of the city. That honor belongs to a similar, but larger, publication known as the Hathaway map, or *Chicago with the school section, Wabansia, and Kinzie's addition*, which was compiled from the original surveys in the Cook County Clerk's office by Joshua Hathaway, Jr. Little is known about this pioneer cartographer except that he arrived in the city penniless and, through a friend, received the job of compiling the map from George W. Snow, the regular surveyor. The publication was sponsored by John H. Kinzie to help the sale of his lots. The detailed correspondence between Kinzie and his agent and partner in New York, Arthur Bronson, reports in 1834 that 600 copies of the map were also printed by Peter A. Mesier, whose firm printed the similar real estate map by John Stephen Wright a bit later that year. The Wright map, it should be noted, includes more city blocks divided into lots, thereby suggesting its later date.

Kinzie's addition to Chicago was plotted in February 1833. Over a year later his New York partner wrote to him that "at length, after importuning the Engraver, Printer, Painter & Bookbinder & Rollermaker," he had assembled a package of 243 maps and was sending them to Chicago. Kinzie had ordered 400 maps, but because the added expense "was trifling," Bronson had increased the edition to 600 copies. He did not indicate on the map that it was a lithograph because he thought that few people would be able to distinguish it from a copperplate work. Six different impressions were made on as many different kinds of paper at a total printing expense of $120.15. Some maps were then mounted on rollers at an additional cost of 75 cents per map. Others were placed in leather cases costing 15 cents each. Most of the maps, however, were sent as individual sheets. Bronson suggested selling the maps for $2.50 on rollers, $1.25 to $1.50 for those in leather cases, and $1.00 for those in sheets. Several copies already had been given away by Bronson who was also involved in the real estate sales. Prominent people in the region like former Governor Lewis Cass received complimentary maps. Several were sent to Washington, and even the surveyor in Cincinnati received a promotional copy.

Peter A. Mesier's lithography firm also produced two other early real estate maps of the city. One, titled *73 Building Lots in Chicago to be sold Auction by James Bleecker & Sons on Thursday, 22nd of October*, appeared in 1835. The lots were located in the school section and the plat was accompanied by a simple inset map of the city locating the area of the lots. Mesier's finest Chicago piece, however, was the large, detailed map prepared by Edward Benton Talcott in 1836. It covered more than twice the area of the 1834 maps, turning the plan horizontally to accommodate all the new subdivisions and thereby putting the lake at the base of the map with north set to the right-hand side. This elaborate production was sponsored by the canal commissioners to stimulate interest in the city and to encourage the sale of their lots in fractional Section 15. Talcott, a native of Rome, New York, came to Chicago in 1835 to work as an engineer on the Illinois and Michigan Canal. Later he took charge of the construction of the canal and also served for a time as the city surveyor. He was active in Chicago as a supervising engineer for a variety of midwestern projects until 1873. He died in Chicago in 1884 where he was eulogized as an original settler of the town and one who had greatly influenced its subsequent development.

Another early real estate map of the city was lithographed by Miller & Company of New York in 1835. The *Map of Lots at Chicago for sale by Franklin & Jenkins on Friday 8th May, 1835 at 12 o'clock at the Merchants Exchange*, is actually two maps of the city. One is a general street map of the center of the city. The other is a detailed plat of a new subdivision west of the north branch of the Chicago River.

The most curious of the early real estate maps of the city was the *Map of the City of Chicago Lithographed & For Sale At N. Currier's Office, Corner of Nassau & Spruce Streets, New York*. It is a very large map with decorative borders covering the area extending from Lake Michigan west to what is now Damen Avenue and from DeKoven Street on the south to Noble Street on the north. In contrast to its elaborate borders, the map is plain and uneventful, printed on dull, blue-gray paper. The diagonal roads were omitted, but a number of public squares was set at regular intervals across the city plan. The central part of the town is filled with numbered lots, but further from the Chicago River these details disappear in the plain blocks stamped with bold numbers. The map apparently was commissioned during the winter of 1836–37 by Amos Bailey, the county surveyor. He employed Asa F. Bradley, a young mechanic, to compile a map which included all the subdivisions recorded up to that time. A great deal of fanciful material seems to have been added to the map to fill up the undeveloped parts of the town.

The Panic of 1837 brought the real estate boom in Chicago to a crashing halt. As the market for real estate evaporated, so apparently did the demand for maps. A decade would pass before another map of the entire city would be printed. Even the chartering of the city in 1837, which divided Chicago into wards, did not produce a

Chicago's First Maps

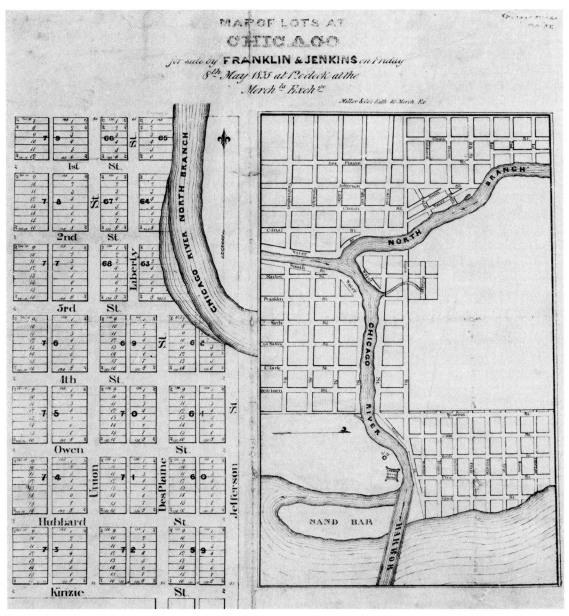

Map published in New York for real estate speculators (1835). A general street map of the center of the city accompanied a detailed plat of a new subdivision west of the north branch of the Chicago River.

21

printed political map showing the ward boundaries. The sole entry of note in the city's cartographic record during this decade is the first surviving map published in Chicago. It was a sketch of Chicago in 1812 to accompany Juliette Kinzie's *Narrative of the Massacre at Chicago*. The pamphlet came from the press of Ellis and Fergus in 1844, but the map also was used a year later by J. Wellington Norris for his *Business Advertiser and General Directory of the City of Chicago for the Year 1845–6*. This slight volume, published by J. Campbell, is probably more worthy of note for including the earliest view of the city from a Chicago publisher, "City of Chicago—South-West View, 1845," which was engraved by Childs & White expressly for the directory.

By 1844 the publication of a directory had become an annual event. As the city continued its growth the need for a map to accompany these directories soon became obvious. The one issued by O. P. Hatheway and J. H. Taylor in 1849 noted that a "beautiful steel engraved map," produced expressly for the book, did not arrive in time for its insertion. The fault, according to the preface, was with the engraver in New York. Whether or not this particular map was ever issued, we do not know. But W. W. Danenhower's directory for 1851 did include a street map of the city, engraved by Doty & Bergen and printed by D. Morse in New York. It was followed in the next year by a similar map accompanying the Udall & Hopkins directory. This map, probably the earliest contemporary map of the city to be printed in Chicago, was produced by Augustus and Charles Burley. After 1852, maps of the city began to be published regularly in Chicago.

These street maps differed in many ways from the earlier real estate maps. They were often smaller in size and emphasized streets, major buildings, and landmarks. They made no attempt to portray accurately the width of the streets. Lot lines and dimensions were never part of the data issued on the street maps. In the late 1840s, however, there was still a need for real estate maps of the entire city. William H. Bushnell's map "exhibiting recent additions, subdivisions, &c." appeared in 1847. Lithographed in New York by Sarony & Major, it was a large cadastral map. A similar *Map of Chicago and Vicinity, compiled by Rees & Rucker, land agents* appeared in 1849. This map, drawn by William Clogher, is unusual because it was lithographed by Julius Hutawa of St. Louis rather than by a New York firm.

The Bushnell and the Clogher real estate maps were among the last of their kind. The new boom in Chicago real estate produced by the completion of the canal and the coming of the railroads created too large of a city to include all the lots on a single map. In a few years, atlases of real estate and fire insurance maps would begin to document the city on a lot-by-lot basis. Map publishing in Chicago soon surpassed that of St. Louis and, a few decades later, New York as well. The culminating publication of the early period of Chicago maps came in 1835. Four maps of the city appeared in that year, and three of them were Chicago imprints. The sole New York entry, drawn by C. Potter, surveyed and published by Henry Hart, and lithographed by Sarony and Company, was the finest map of Chicago published in the early period. Its huge size permitted a detailed representation of the entire city. As in the early real estate maps, it showed some land ownership in the larger tracts. Every lot was shown according to scale. It was also a street map of the entire city and included hundreds of features like buildings, factories, railroads, and parks. It was one of the earliest maps to indicate wards by separate coloring. Its fine printing included a vignette of the courthouse. Thus it summed up the first several decades of mapping Chicago, soon to be the city of maps for the entire nation.

For Further Reading

THE STANDARD LIST of early maps of Chicago, prepared by Marsha Selmer, appears in volume 4 of Robert W. Karrow, Jr., ed., *Checklist of Printed Maps of the Middle West to 1900* (Boston: G.K. Hall & Co., 1981). Several of these early maps have been reprinted in modern editions, but all lack critical introductions. Paul Angle's "The First Published Map of Chicago," *Chicago History* 3 (Spring 1952), is on the Hathaway map. See also the correspondence between John H. Kenzie and Arthur Bronson found in the manuscript collection of the Chicago Historical Society. The Story of John Wright can be pieced together from Lloyd Lewis, *John S. Wright, Prophet of the Prairies* (Chicago: The Prairie Farmer Publishing Company, 1941) and the promoter's own voluminous writings. One of the earliest references to map publishing in Chicago is found in the *Annual Review of the Commerce...of Chicago for the Year 1854* (Chicago: Democratic Press Job & Book Steam Printing Office, 1855). Grant Dean, map curator at the Chicago Historical Society, is especially knowledgeable about early maps of Chicago.

Rufus Blanchard: Early Chicago Map Publisher

By Marsha L. Selmer

For half a century beginning in 1854, the name Rufus Blanchard was synonymous with Chicago mapmaking. His career began at the Chicago Map Store at 52 LaSalle Street.

ALTHOUGH NOT WIDELY KNOWN today, Rufus Blanchard was a major contributor to the nineteenth-century Chicago mapmaking scene. During his half century in Chicago from 1854 to 1904, he published maps of Chicago, all of the north central states, Texas, and the regions of the United States. In addition, he produced both outline and history maps for classroom use. Because he focused on map rather than atlas production, much of his work has gone unrecorded. In the absence of business records, it is difficult to determine the extent of his creative role in producing the maps he published, and therefore his publishing career must be reconstructed from his maps. The autobiographical sketch which appears in his *History of DuPage County, Illinois* suggests that he operated largely as a publisher. An examination of his primary imprints indicates, however, that he was also a compiler and an engraver at the peak of his cartographic production in Chicago. As used here, "primary" imprints denote those in which the sole or principal cartographer and publisher is Blanchard, or Blanchard along with a partner. "Secondary" imprints are works of other cartographers published by Blanchard, or in which his name appears as either co-publisher or in a subordinate position.

The Early Years

Blanchard was born on March 7, 1821, in Lyndeborough, New Hampshire, one of six children born to Amaziah Blanchard and his first wife, Mary D. Blanchard. By his own report he left his New England home in 1835 for New York City. How he occupied his time there, whether by pursuing a trade or furthering his education, is unclear. His name was not among those listed in the city directories of the period, but it is probable that Rufus made his home with his brother Calvin. In 1837 he moved to the Ohio frontier and spent the next few years farming, hunting, trapping, and teaching. By 1840 he returned to New York where apparently he was employed by Harper Brothers as a salesman for their publications, though no Harper Brothers records survive to support this claim.

Blanchard opened a bookstore in 1843 in Lowell, Massachusetts. While in Lowell, he boarded with his sister Anna, brother-in-law Joseph, and their infant son George Franklin Cram. During 1846 he moved his bookstore to Cincinnati where he employed his brother, Edwin A. Blanchard, as a clerk. It is clear that in Cincinnati he sold maps as well as books, and by 1848 he was adding his name as retailer to the maps that he stocked. During this period he opened a branch store in New Orleans, seeking to tap further the growing markets of the West and South.

Returning to New York in the early 1850s, he continued in the book trade and authored and copyrighted a chart titled *The Grammatical Tree, Showing the Classifications and Properties of the English Parts of Speech* (c. 1853), which probably was intended for classroom use. He also became acquainted with Charles Walker Morse and was exposed to the techniques of cerography, or wax engraving, which would become central to his mapmaking career.

Chicago's Earliest Map Publisher

In the early 1850s Chicago became the focus of railroad links between the East and the beckoning West, something that made it a favorable site for the beginning of Blanchard's new endeavor. Maps for immigrants, real estate promoters, shipping companies, and railroads were obviously

---

Marsha L. Selmer is map librarian and assistant professor at the University of Illinois at Chicago.

Early Map Publisher

Early advertisement for Rufus Blanchard's Chicago Map Store at 52 LaSalle Street. Before publishing maps of his own, Blanchard retailed those of other firms.

the order of the day. In 1854 Blanchard, then thirty-three years old, opened the Chicago Map Store at 52 LaSalle Street. The store's logo (the Western Hemisphere of the globe), which appeared on some of his early map and newspaper advertisements, identified him as a map publisher. The transition from New York to Chicago is shown by *Morse's Map of Illinois* (1854) and *Morse's Cerographic Map of Wisconsin* (1855), which Blanchard published in Chicago but copyrighted and printed in New York. Other advertisements indicate that Blanchard sold maps, books, and prints, both wholesale and retail. According to an R. G. Dun & Co. credit report from June 26, 1856, Blanchard carried Chicago's largest stock of maps and clearly was prospering. By 1857, he had begun to publish maps that were his own work, including his *Map of Chicago*.

We know little of his personal life but that his first marriage ended tragically when his bride died of injuries from a railroad wreck on their honeymoon. Soon thereafter, in 1858, Blanchard launched the publication of the *Northwestern Quarterly Magazine,* which failed with the first issue. His financial loss in this venture was substantial. During the next few years credit reports characterized his map store as barely breaking even, and his cartographic publications ceased altogether. In spite of his setbacks, or perhaps because of them, he married again about 1860. His new wife, Annie E. Hall, twenty-one years his junior, was engaged primarily in homemaking but traveled widely with her husband over the years, helping him gather the material he used in his publications. Blanchard resumed map publication during the Civil War when business prospects improved.

The Cartographic Peak: 1865–1877

In 1865 the geographical scope and the quality of Blanchard's maps began to increase. He operated totally on his own, though he sometimes contributed to the maps of out-of-town publishers. There are only four instances of short-lived publishing partnerships. The first of these was the most significant.

In 1865 Blanchard's nephew, George Franklin Cram, was discharged from the Union Army and returned to the Chicago area, where he apparently entered into a traditional two-year apprenticeship with his uncle. During 1867 and 1868, they published maps that were clearly new editions of previous Blanchard efforts. They also apparently altered and shared the same printing plate for the 1868 *Guide Map of Chicago,* which each man issued independently. The Cram version, issued by Cram and published by lithographer Charles Shober, contains the following four lines, which appear below the main title and partially overlay the water lining:

> Published by Chas. Shober & Co.
> Engr and chemityped by
> Shober
> Chicago

On the Blanchard publication only a faint shadow of these lines is visible on the water lining below the main title, indicating that in the beginning

---

Opposite: *Morse's Illinois map from 1854 is an example of Blanchard's practice of publishing the work of other cartographers, in this case Charles Morse. By 1857, he had begun to publish his own maps.*

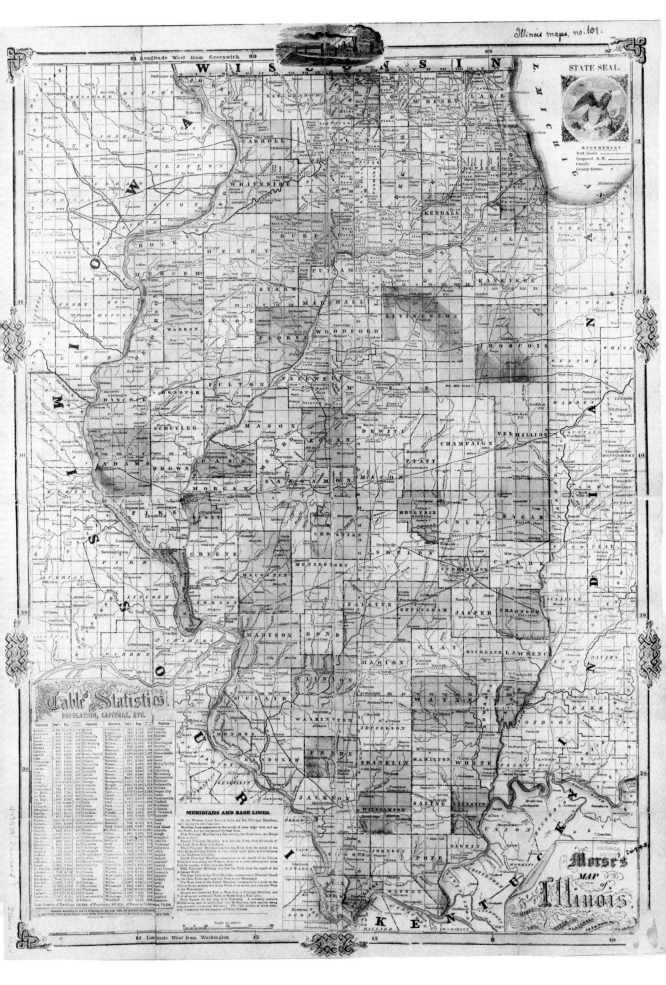

Early Map Publisher

Sheet maps were, among other things, vehicles for trade advertising, as in this border from Morse's map of Illinois. New advertisements might appear with each new printing.

Blanchard most likely published maps with Cram to establish his nephew's credibility in the Chicago map trade. A partial division of the products was also one way to reduce direct competition between them and give each an opportunity to prosper. Such a division of labor is mirrored in their subsequent activities; nearly all of Blanchard's cartographic works were maps, and Cram's were atlases.*

While Blanchard claims to have "introduced the manufacture of maps in the West, in all its departments," his personal role in making the maps he published remains elusive. On some of his maps published from 1868 to 1874, he identifies himself as compiler or engraver, but on the majority only as publisher. The oval-shaped advertisement which appeared on many of his maps and on his business stationery during the 1870s reads: "Map drawing, engraving printing, coloring, and mounting executed in the best style, Rufus Blanchard." Whether Blanchard did all this work himself on any particular map or spread the work among employees and subcontractors cannot always be determined. Since he traveled widely to gather the information which appeared on his maps, it is probable that he was responsible at least for compiling the initial manuscript.

It is clear that Blanchard dealt with a number of other Chicagoans in the business of map production. Several of the sheet maps that he published between 1863 and 1870 were lithographed by Shober, a general lithographer who worked for a number of mapmakers, including Cram and Alfred T. Andreas. Others whose names appear on Blanchard's maps during this period as either draftsmen or engravers are Otto Peltzer (1867–68) and Jerome T. Gouget (1869). Mary E. Smith, who was "map finisher" for Blanchard in the late 1860s, worked as a map mounter throughout her career and was the only woman in the Chicago map trade who was not the widow of a mapmaker. Anders M. Askevold was a map engraver with Blanchard in 1871.

While his business in Chicago prospered, Blanchard moved his residence to Wheaton, Illinois, where he would remain for the rest of his life. There in the autumn of 1871 he established a "map factory," which was eventually housed on the grounds of his home. That October along with thousands of others, Blanchard suffered devastating losses in the Chicago Fire. A note handwritten by Blanchard in 1890 and attached to a copy of *Blanchard's Map of Cook and DuPage Counties* (1871), states simply: "This map was published just before the great fire of 1871. Nearly the entire edition was burned."

Blanchard's second publishing partnership, with Zebina Eastman, was launched in 1872 at 132 Clark Street, where it remained for about five years. During this period, only the partnership of Eastman and Bartlett, booksellers and stationers, is evident in directories. Although Eastman began his career in Chicago as a printer, he was better known as an editor and publisher. Since the partnership was established in the year immediately following the Chicago Fire of 1871, it is possible it was undertaken to help Blanchard through a financially tight period. But neither the Great Fire of 1871 nor the nationwide panic of 1873 resulted in a significant decrease in

---

* In *The All-American Map,* David Woodward writes that Blanchard died in 1867, whereupon Cram took over the business and renamed it the George F. Cram Company. Based on typewritten notes regarding the company's history obtained from the George F. Cram Company, now in Indianapolis, that evidence is inaccurate and misleading since Blanchard lived beyond 1900.

Early Map Publisher

Blanchard's output of maps.

Blanchard published maps of all areas of the United States on either a state or regional basis. His state and regional maps generally included railroads, both those proposed and built, township and range divisions, and populated settlements. As the midwestern and Plain states grew, Blanchard and his competitors produced exceptionally detailed state maps that even portrayed individual 640-acre sections. His maps of Chicago appeared regularly, and when a major annexation took place in 1869, he published two editions in the same year. Estimates of Blanchard's cartographic output between 1854 and 1904 show that 1870 was his most productive year.

The advertising lists of Blanchard's stock, which appeared in his works from 1867 through 1893, provide some clue to the format, price, and scope both of his own maps and others that he sold. Prices varied according to the scale of the map — the more detailed were more expensive — and the form in which it was sold. Maps could be bought in flat sheet form, half-mounted, mounted on rollers, folded in pocket form, on cloth, and in morocco case. He sold his competitors' maps at a higher price than his own. For example, in 1867 Blanchard's map of Illinois and parts of adjacent states, at a scale of twenty miles to an inch, sold for 35 cents in sheet form, 50 cents in pocket form, and 75 cents when mounted. Blanchard's township map of the state of Illinois, at a scale of twelve miles to an inch, sold for 75 cents. Farmer's township map of Michigan and Colton's township map of Iowa each sold for 75 cents, while Chapman's sectional map of Wisconsin and the sectional maps of Colton and Farmer sold for $1 to $1.50 in pocket form. Maps by Rice, Bancroft, and Thayer are also listed in his stock during 1873.

The inclusion of Blanchard's imprint on the maps of others indicates that they too were a part of his sales stock, although not all were listed in his advertising. Maps of this type include A. J. Johnson's 1858 map of Michigan; C. W. Morse's 1855 map of Indiana; George O. Willmarth's 1865 map of Kansas; Silas Chapman's 1865, 1866, 1867, and 1868 maps of Minnesota; H. H. Lloyd & Co.'s 1867 map of Ohio; J. H. Colton's 1871 map of Illinois; and G. W. and C. B. Colton's maps of Illinois, Iowa, and Missouri.

Although Blanchard was aware of the wax engraving technique and had begun his Chicago career publishing the cerographic maps of Charles W. Morse, once he became established in Chicago he generally used lithography. The back cover of his 1867 *Handbook of Minnesota* includes a description of the methods used at this time. "The first thing to be done is, to draw the map by hand exactly as it is to appear when printed." First the base map was compiled from the data found in the original government surveys, then county boundaries and town locations were obtained from each county, and finally, information on railroads and railroad stations was gathered from the different railroad companies. The completed manuscript map was then engraved on a copperplate or a lithograph stone. After printing on a lithograph press, colors were applied by hand. Pocket maps were printed on bond paper and folded into muslin cases. Wall maps were mounted on stretched, bleached muslin, then varnished and tacked onto rollers.

Blanchard's first attempt at atlas production also appeared during the 1870s and was based on maps issued initially in sheet form. His *New Commercial Atlas of the United States* (1874) with a title page, table of contents, and pagination, meets the criteria for an atlas. The pagination, however, was added by rubber stamp.

During his travels throughout the Midwest

gathering information for his maps, Blanchard developed an interest in American history, and in 1876 published his *Historical Map of the United States*. Sectioned and mounted on boards, the map could be hung from the wall or folded into covers. A tablet of history was mounted on the verso. The local press quickly took note and the February 16, 1876 issue of the *Northwestern Christian Advocate* even claimed incorrectly that Blanchard was the "first one to inaugurate this system of showing history on maps." Emma Hart Willard first introduced this kind of map in 1828 with *A Series of Maps to Willard's History of the United States, or Republic of America*.

A New Direction: 1878–1904
Blanchard's pursuit of history culminated in a simultaneous career as a mapmaker and an author of American history. From 1878 to 1903 he published his own works of history and patriotic verse. The best-known work is the *Discovery and Conquests of the Northwest*, which was published in three editions. The significance of this history today lies in its reproduction of primary material.

Blanchard was to suffer one more business and financial disaster in his career. On April 22, 1885, a fire swept through Wheaton, this time shortly after the morning commuters had departed for Chicago. Many homes and businesses were destroyed. Blanchard's home and furnishings were saved, but his wife was able to rescue only a few volumes from the building that housed his personal library, his stock of books and maps, and his map plates.

From 1878 to 1886 the few cartographic works that Blanchard created, whether sheet maps or maps in text, dealt with the history of Illinois. Beginning in 1887 his map productions resumed, but for the remainder of his life his cartographic output was focused on primary imprints of

Early Map Publisher

Above: *One of the first figures in the Chicago map trade, Rufus Blanchard published maps in Chicago for half a century after his arrival in 1854.*

Left: *Cartouche from Blanchard and Cram's Illinois township map (1867). Meant for commercial use, the map meticulously charted the railroad network and located every city and town.*

Illinois and Chicago and its suburbs. This narrowed geographical scope and reduction in the number of Blanchard's maps issued annually, was the result of his growing interest in writing and publishing history, increased activity among other mapmakers (particularly his nephew George Cram), and the reduced rate of regional settlement change in the country.

Blanchard also attempted publishing another serial in 1888, *The Map Graphic*. It comprised both maps and text devoted to the history of world exploration, the history of Chicago, historical maps of the world, contemporary maps of Chicago, and mapmaking techniques. But it was short-lived, and only two issues for January and April, in a newspaper format of four pages each, ever appeared. Despite this setback, Blanchard soon returned with another effort in December 1889, a one-sheet "newspaper" advertising Chicago's 1893 World's Columbian Exposition, which he edited and published in partnership with Isaac H. Taylor.

By the 1880s Blanchard had returned to the use of the wax engraving technique. During this time and the early 1890s, some of Blanchard's maps were drawn by B. F. Davenport (1883, 1886, 1888), and others were engraved by A. Zeese & Co. (1888, 1890, 1892). In addition, his maps were also issued by other publishers such as the National School Furniture Co. (1883) and Rand, McNally and Company (1895). Both were also publishers of Blanchard's books, *Rise and Fall of Political Parties in the United States* (National School Furniture Co., 1884) and *History of Political Parties in the United States* (Rand, McNally and Company, 1884). The texts were actually the same but the Rand, McNally edition included one of their maps after the title page.

The remaining atlases credited to Blanchard date from the mid-1890s and, like earlier editions, were created from separately issued sheet maps. The *Historical Atlas of the United States* (1895), and the *Atlas of Chicago* (1895–96) are examples. The first of these has a title page that includes a table of contents and pagination added after the maps were mounted. The second has none of these features. Thus while Blanchard was associated with works loosely called atlases, he was not actually an atlas publisher.

Blanchard's final publishing partnership, with Arista C. Shewey, involved the production of bicycle maps, popular at the turn of the century because of the widespread interest in this form of recreation. The partnership is not evident in city directories, and both were listed as separate business addresses at this time. Shewey also engraved and published maps independently during the 1890s.

In later years of his life the personal role that Blanchard played in the creation of his maps was severely limited. The letters he wrote during the late 1890s and later, by the time he was in his seventies, were either dictated or typed due to his "shakey hand." Clearly, this physical problem would have precluded his drafting or engraving the maps he published during this period. Nonetheless, he continued active work. An article in the September 8, 1903 *Chicago Tribune*, just four months before his death, credits him with introducing the mapmaking industry to Chicago and with being the oldest map publisher in the United States. He was reported to be at his Randolph Street office every day. Blanchard died at the age of eighty-two on January 3, 1904 in Wheaton, Illinois. His wife retained his maps,

The first issue of Blanchard's short-lived periodical, The Map Graphic, *from January 1888. Although only two issues ever appeared, Blanchard remained a major force in Chicago mapmaking until his death in 1904.*

books, and printing plates, valued at $211.60, leaving the balance of the small estate to other family members.

The Blanchard maps published posthumously can be attributed to his estate, which is listed in all of the Chicago city directories at least through 1917. The estate was managed by his nephew George Cram, and was located at Cram's business address. Annie Hall Blanchard died on September 27, 1925. Probate records relating to her estate, which might have offered a clue regarding the disposition of her husband's business records, have not been found. As a corresponding member of the Chicago Historical Society, he made gifts and had business dealings with the institution, although his records were not placed there. It is possible they were kept by George Cram, as manager of the estate, but they are not in the Cram Company archives today.

Blanchard's fifty years as a map publisher spanned a period of major change in cartographic production techniques. He had started his Chicago publishing career with maps produced by the wax engraving technique and they had sold well. Why then did he turn away from this technique and use lithography at the height of his productivity? And what prompted him to return to the wax engraving method in the last twenty years of his career?

The reasons for the change in his production style at various points in his life can be inferred from the article on mapmaking he published in *The Map Graphic* of 1888. Lithography, which had been in use prior to the invention of wax engraving, was an expensive method of engraving high quality maps. The printing stage was slow and expensive. While the engraving phase for wax engraving was more expensive than that for lithography, the savings realized in printing from a wax engraving made it a more economical technique for producing large editions of maps. But, as Blanchard tells us, large editions were not always the most profitable:

By this time the building of railroads was begun and it became necessary to publish new editions of all these maps often to keep pace with the new towns that were constantly springing up along new lines. To correct these metallic plates was difficult and expensive, far more so than to make corrections in lithograph and under this strain, the new style was abandoned and lithography substituted in its place as the most practical way of mapping a country to which the flood gates of emmigration were thrown open.... The number of maps wanted now [is] greater than ever before and how to keep them corrected in metallic plate is the problem to be solved. After many years of experience the writer has formulated a plan by which this can be done with as much facility as if the map was made by lithography and future maps published by him will appear in the new style original with him and new editions of them, neatly corrected, will be issued as often as necessity may require. Let it not be understood that... lithographic maps will ever go out of use. Small editions of [lithographic maps] can now be made cheaper than by any other process....

The changes therefore were determined by economic factors rather than the presence or absence of skilled craftsmen in Chicago. The introduction of the steam lithographic press in 1865 also may have contributed to his use of lithography.

Throughout his career Blanchard was a regional publisher, issuing maps covering Chicago, and combining his own work with that of others in issuing maps of Illinois. At the peak of his cartographic productivity, between 1865 and 1877, he also published maps of the individual states and regions of the United States for use by a wide public. From the late 1870s onward, his maps were often issued for a specific purpose, for a book, for bicycling, or for classroom teaching. These published cartographic works stand as an impressive record of Blanchard's half century as a Chicago mapmaker and his role in establishing the Chicago map trade.

For Further Reading

THE EVIDENCE for reconstructing Rufus Blanchard's career is scattered. A useful, though brief, biographical sketch appears in his *History of DuPage County, Illinois* (Chicago: O.L. Baskin and Co., 1882). Some family information can be gleaned from the manuscript schedules of the 1870 and 1900 U.S. Population Census, probate records held at the DuPage County Court in Wheaton, and interment records at Graceland Cemetery, Chicago. Other manuscript material includes limited business data in the credit reports of R. G. Dun and Co., Illinois Volume 27, p. 326, held by the Baker Library, Harvard Business School, Boston, and some autograph letters at the Chicago Historical Society. References to Blanchard's activities also can be found in city directory listings for Chicago, Cincinnati, New Orleans, New York, and Lowell, Massachusetts, as well as newspapers such as the *Chicago Daily Tribune*, *Chicago Tribune*, *Herald Record*, *Wheaton Illinoian*, *Wheaton Progressive*, and Blanchard's own *Map Graphic*. Blanchard's extensive map output was examined primarily at the Chicago Historical Society and also at the Library of Congress.

George F. Cram and the American Perception of Space

By Gerald A. Danzer

For Americans who craved information and liked to keep up-to-date, George F. Cram offered fact-filled and affordable atlases that portrayed an orderly though changing world in which all the pieces fit together as progress marched ahead.

THE NAME George F. Cram has been identified with the making and selling of maps for more than a century. For half that time the name was identified with Chicago. Recognized in his own day as an astute businessman, it is now time to put his career as a map publisher for the popular audience into a broader perspective. His maps and atlases have more than simply biographical significance: they also offer some important clues about the American perception of geographical space in the late nineteenth century.

Two incidents highlight the importance of Cram and his maps in the emerging city and in its commercial hub. When the West Chicago Park District was established in 1869, the first thing the commissioners purchased were certified copies of the act which brought them into existence. Their second purchase was a map of the city from George F. Cram's store. On that occasion, Cram may have sold them a map published by someone else. A decade later, however, he had more to offer. For private business too, Cram's maps became an essential fixture. The first page of the journal of the Lewis Publishing Company, dated November 12, 1880, for example, listed as "expenses" the items necessary to begin business: a table, a desk, a bookcase, an office chair, two cane chairs, a pigeonhole case, an oil can, Webster's dictionary, window shades, an ink stand, company stationery, and one of Cram's map.

Cram's map cannot be identified with precision. George Franklin Cram had published dozens of maps by 1880. Perhaps it was an atlas rather than a single map. Cram's earliest atlas of record, *The New Commercial Atlas of the United States* (1875), would have been an appropriate choice. Another publication, *Cram's Indexed Commercial Atlas of the Western States*, subtitled "A Complete Guide for Business Men," appeared in 1879, just before the office was furnished. The following year saw an early version of what became Cram's most celebrated publication, *The Standard Atlas of the United States*. This inexpensive version evolved into the *Unrivaled Family Atlas* series. The simple statement, "The 64th edition," on the 1952 issue is reason enough to include George Cram in the history of American cartography. "Any map," in the choice words of map historian R. A. Skelton, "is a precipitation of the spirit and practice of its time." This is equally true of Cram's atlases, volumes which provided simple and connected pictures of the world through maps.

Cram's Map Career

George Franklin Cram was born on May 20, 1842, in Lowell, Massachusetts, the son of Joseph T. Cram and Anna D. Blanchard. A year later his uncle, Rufus Blanchard, opened a bookstore in Lowell and began a career in maps and books that he would follow for the next six decades. Blanchard soon moved his business west, first to Cincinnati and then in 1854 to Chicago. By the early 1860s, Rufus Blanchard had developed a successful business publishing a few maps and selling a combination of maps, globes, and books. He made his home in Wheaton, Illinois, where he apparently served briefly on the faculty of the Illinois Institute, the predecessor of Wheaton

---

Gerald A. Danzer is associate professor of history at the University of Illinois at Chicago.

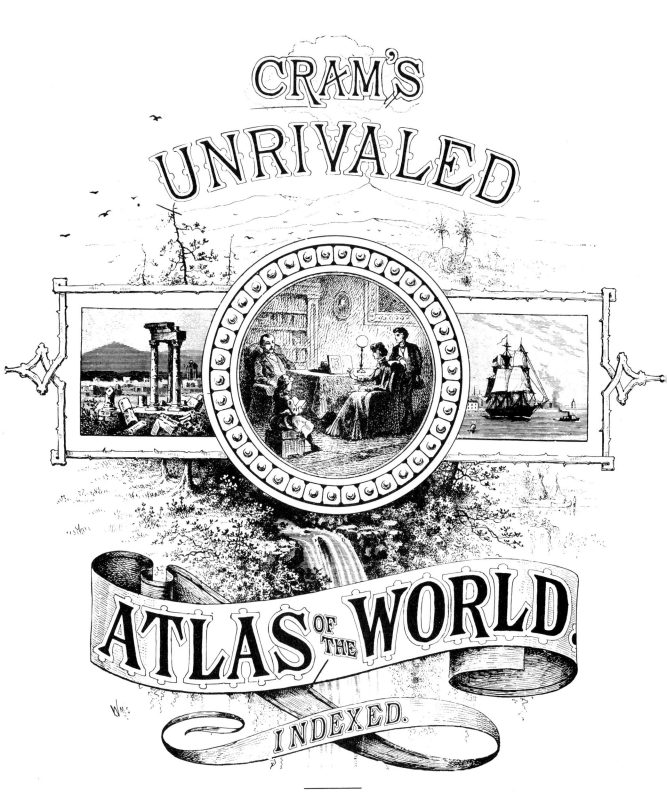

The title page of Cram's Unrivaled Atlas *(1889) graphically summed up the connections forged by an atlas between small worlds and large ones. The family learned about the world and their place in it through the atlas, which carried them, imaginatively, to thousands of distant and exotic places.*

American Perception of Space

Among Cram's state maps were those laden with statistical information, as in this example showing Illinois farm and crop values and agricultural products by county (1906).

College. It is logical to assume that George F. Cram came to attend Wheaton College on the advice of his uncle.

The school catalog for 1860–61 lists Cram along with fourteen other gentlemen and one lady in the senior preparatory class, but early in 1862 he must have dropped his studies in favor of the Union cause. The list of students in the Wheaton College catalog grew shorter as the war years rolled on. By 1863 there were only two students left in Cram's class, a mute testimony to the war's impact. A first sergeant in Company F of the 105th Illinois Volunteers, Cram served in the Union Army until the end of the war, seeing action in the battles of Resacca, Beach Tree Creek, and Chattanooga.

It is impossible to know what influence this military experience had on the young soldier. Common sense indicates that three years' service during his formative years must have had a substantial impact. Hints of this can be found in the only two publications, other than maps, atlases, and geographical handbooks, that came from Cram's own pen during his lifelong association with books and publishing. The first, *Pensions: Who are Entitled to Them and How They May Be Obtained* (1886), was dedicated to "every soldier, his wife and children, to every soldier's widow and orphans...in the hopes that...all will get their just desserts." Cram apparently compiled this collection of statutes, regulations, and general information in association with R. A. Tenney, another Chicago mapmaker. The second, Cram's historical novel entitled, *Minette, a Story of the First Crusade* (1901), combined a love story with military adventure.

During the war, Cram may have perceived a business opportunity in the public demand for detailed maps of the various battle theaters. Perhaps he had become familiar with cheap, popular publications such as *Philips & Watson's Historical and Military Map of the Border & Southern States*, published by the New York firm in 1863 with an accompanying pamphlet giving brief descriptions of battles and skirmishes. Printed on some copies of the map just below the title was the announcement, "Agents Wanted for our Maps and Charts." The back cover of the pamphlet also advertised the publishers' "General Depot for Agents to get their stock of the most popular articles for the Times."

On October 3, 1865, soon after Cram had returned to Wheaton from the army, he married Martha A. Hiatt, a local resident. They first lived in Wheaton, while Cram was employed as a clerk in Rufus Blanchard's firm. In 1867 the household moved to Evanston, and Cram began to move upward in the business world. The same year found him a partner in the firm of Blanchard & Cram, and two years later, in 1869, he established his own firm, George F. Cram & Co. Cram also had several business associations with Watson's Map Depot in Chicago. When Cram first entered business on his own in 1867 or 1868, he set up an own office at 134 Lake Street. The next year's city directory listed him as the proprietor of the Western Map Depot. Another source lists this establishment as Watson's Western Map Depot at 66 W. Lake Street, just down the street from Cram's first office. After the Great Fire wiped out the entire business district in 1871, the firm reappeared as the Cram Map Depot at 55 W. Lake Street. Obviously there was a close connection between the Cram and Watson business interests in Chicago. Between 1868 and 1885 the Crams made their home in the Union Park area, about a mile and a half west of the central business district. In 1879 they conducted business at 264 Wabash and lived in a two-story frame house at 841 W. Washington Boulevard, an upper middle-class area of the city. The credit reports in the records of R. G. Dun & Co. document the achievements of the thirty-eight-year-old businessman. He owned his home free and clear, he had money in the bank, and his business was profitable.

It was during this period that Cram began publishing the atlases that were the key to his further business success. In 1885, the very year in which the *Unrivaled Family Atlas* reached its tenth edition, Cram's name appeared for the first time in *The Elite Directory and Club List*, a social register. The following year the family moved to 4168 Drexel Boulevard, in the fashionable district of the city's South Side. By 1890 a daughter and her husband, John W. Iliff, moved next door. The Crams moved in with the Iliffs in 1896, perhaps because of Martha Cram's failing health. But George Cram's business activities continued to expand. In 1898 he joined his aged uncle in publishing *Blanchard's Chicago Guide*, by then a popular standard map of the city. This association continued until Blanchard's death in 1904, when Cram took over full managerial responsibility for Blanchard's business and publications. Two years later the familiar Chicago map became *Cram's Street Guide*, a publication that continued until the late 1930s, a full decade after Cram's death.

Martha Cram died on August 7, 1907, and three years later George married Leonia Wilson. They lived in the house at 4166 Drexel Boulevard, apparently continuing for a time to share the residence with the Iliffs. Cram retired in 1920, selling his business the following year to the National Map Company of Indianapolis, a firm with whom he had done business for a number of years. He died on March 24, 1928, at the age of eighty-six. The *Chicago Tribune*, in announcing Cram's death, referred to him as a "wealthy retired publisher and traveler." He was buried in the Blanchard family plot in Graceland Cemetery in Chicago.

The Cram Company

Throughout the early years of his career, up to the time of the fire, the young Cram was largely a free-lance entrepreneur, joining with his uncle and others in various publishing and merchandising ventures. His earliest surviving imprint, *Guide Map of Chicago*, was one of six city street maps published in 1868 with identical or similar titles. A careful comparison of them reveals something about map publishing in the 1860s.

All six maps are actually the center part of a larger piece, *Blanchard's Map of Chicago and Environs*, published the same year. This larger map was lithographed by Charles Shober from a draft by Otto Peltzer. Such cooperative efforts were frequent among Chicago mapmakers, and each individual associated with this particular map became, in his own right, a noteworthy figure in the history of Chicago cartography.

A single map often was used for a variety of "jobs." Each impression for a particular purpose frequently would carry its own special imprint, listing the individuals responsible for the specific undertaking. Each impression might, therefore, have a different publisher. Hence, identical copies of the *Guide Map of Chicago* appeared under the imprints of Rufus Blanchard, Blanchard & Cram, Chas. Shober & Co., and George F. Cram. Blanchard's issue was folded into a twelve-page pamphlet with orange paper covers and titled *Guide Map of Chicago*. Blanchard also sold some of these maps to John S. Wright to be bound with copies of his book, *Chicago: Past, Present and Future*, published in Chicago in 1868. The Blanchard & Cram version appeared in a more elaborate format: a twenty-four-page *Citizen's Guide for the City of Chicago* with cloth cover stamped in gold. The Cram edition appeared in a sixteen-page guidebook and also was issued separately. In the latter map, George F. Cram is listed as the publisher, but additional type under the date added: "Published by Chas. Shober & Co. Engr. and chemityped by Shober, Chicago." Blanchard & Cram also revised portions of the map, added some views of the city, and sold this new version as a separate publication.

Such multiple use of map material well illustrates the *ad hoc* nature of Chicago's early map business. There is little evidence that Cram, like the others, had any large, comprehensive production plans. He made partial investments in small-scale projects which, if successful, might supplement the major purpose of the firm: the selling of maps and other goods. In the absence of business records, it is easier to trace Cram's publishing activities than his career as a map seller. Many of the maps have survived, but there are no catalogs or sales records, and only a few advertisements. Yet merchandising was probably Cram's major source of income in the early years. Most of the maps he sold were published by others, and maps were only one part of his store's stock. In merchandising, as in map publishing, it is difficult to determine exactly who owned the business. The city directory of 1866 gives Cram's occupation as a clerk while he worked for his uncle. The 1868–69 edition lists George F. Cram in the classified section under "Map Publishers and Dealers," with an office at 134 Lake Street, just a few doors away from Blanchard's at 146 Lake Street.

The business section of the following year's directory continued to list Blanchard at 146 Lake, but James Van Vechten was given as the name of the firm at Cram's old address, 134 Lake Street. Moreover, in the general directory of names, G. F. Cram & Company was listed at the same address in some kind of association with Martin Griswold. In 1870 Van Vechten was back at his old Clark Street address, and George F. Cram & Company was now next door to Blanchard at 146 Lake

American Perception of Space

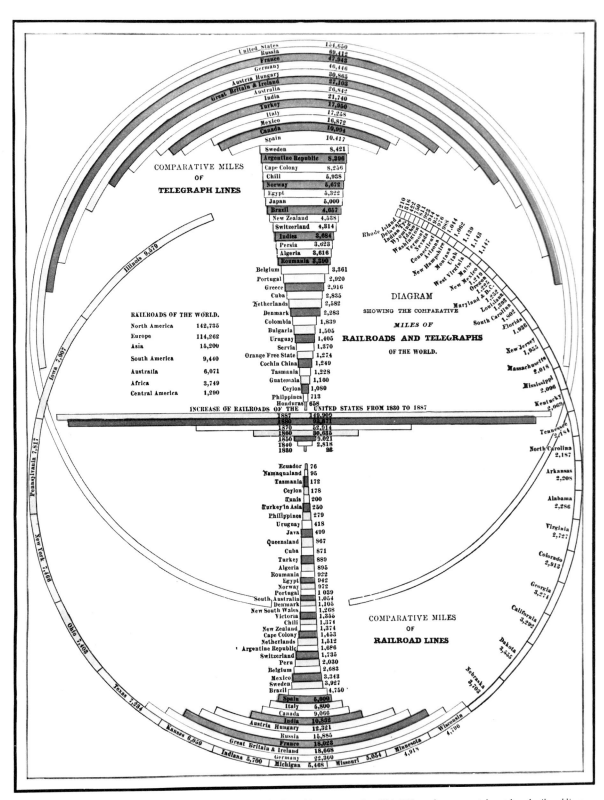

In addition to maps, Cram's atlases commonly included other kinds of graphic representation. This 1889 graph compares telegraph and railroad lines worldwide. In both categories the United States led the world; in railroad mileage Illinois led the nation.

Street, sharing the building with another map publisher, J. W. Goodspeed & Company. The 1871 directory noted several more address changes for map dealers and listed a new addition, "Watson's Western Map Depot, D. Needham, manager, 66 Lake." Cram's 1869 Iowa map listed Cram as the proprietor of the Western Map Depot. At any rate, the Great Chicago Fire of October 1871 simplified things, destroying the entire stock of every map dealer in the city, and driving several firms out of business.

Cram responded to the disaster with characteristic energy. With the ashes still warm, he produced a lithograph of the event. "The Destruction of Chicago, 8th, 9th, & 10th of October, 1871" was advertised as "the first picture published in Chicago after the fire." It carried the imprint "G.F. Cram, 55 West Lake St., Chicago," but also advertised a Cram Map Depot. The view is from Lake Michigan extending from Lake Street to Jackson Boulevard. The trains lined up in the foreground seem out of place, implying that it may have been an adaptation of an earlier print. Its crude quality, almost like a woodcut, suggests a hasty production.

A special Fire Edition of *Edward's Chicago Directory* listed the relocation of various firms as the central business district was being rebuilt. Cram inserted a special advertisement in this guide:

George F. Cram
Proprietor of the
Great Western Map & Picture Depot
55 W. Lake St. Chicago, Ill.

In succeeding directories Cram dropped references to the map depot and listed his firm simply as George F. Cram. In 1873 he moved down the street to 66 W. Lake Street, remaining there until 1879, when the firm moved to Wabash Avenue. In 1888 Cram moved once again, this time to 415 Dearborn, and a decade later to 61 Plymouth Court. In 1906 the firm expanded to 552-54-56 Wabash. Finally in 1913, it moved to its last Chicago location at 109 N. Market. The periodic moves became necessary as the firm grew in size and as the nature of its business changed. The early locations on Lake Street were in the retail sales area of the city. As publishing projects expanded, retail sales grew less and less important. The locations on Dearborn, Wabash, and Market streets all were in areas where printing and publishing companies were concentrated.

After 1868, Cram left the publishing of Chicago guide maps to his uncle, issuing only occasional local maps while Blanchard was in business. Instead, Cram turned to state maps of Iowa, Illinois, Missouri, and Kansas. In 1874 he added maps of Wisconsin, Colorado, Indiana, and Minnesota. By 1879 he was able to combine these various state maps into *Cram's Indexed Commercial Atlas of the Western States: A Complete Guide for Business Men*. Cram's entry into the atlas field actually came several years earlier. The *New Commercial Atlas of the United States and Territories...A Complete Shipper's Guide* appeared in 1875 under the imprint "New York and Chicago: The Cram Atlas Company." This eighty-two-page volume actually was made from a wall map of the United States that had been cut up into various state and territorial maps. Every other page carried a dozen or so advertisements, almost all of them for Chicago firms. Printed in Chicago at the Lakeside Press, the atlas appealed only to a local audience. Thus, the reference to New York on the title page is something of a mystery.

A year later in 1876, Rand, McNally and Company, Cram's rival, issued its first and more successful *Business Atlas*. Soon both firms were publishing a variety of atlases in Chicago. Although Cram's atlases matched Rand, McNally's in coverage, in general, they were not of equal quality. Because he aimed his atlases to reach the widest audience possible, Cram settled for cheaper and thus lower quality books.

Cram's *Standard American Atlas of the United States* (1880) had only forty-one maps and eighty-six pages. A year or two later he issued several versions of a world atlas as well as an *Illustrated Handbook of Geography*. Soon both the atlas and the handbook of geographical information were combined in a single volume, creating a large, imposing atlas that became popular with the general public. Cram saw the advantages of publishing atlases in several sizes and styles to meet a variety of needs. Between 1880 and 1920 the firm published several hundred separate atlases, ranging from ponderous reference volumes to inexpensive pocket versions. The titles of these various publications suggest the range of the market: *The People's Illustrated and Descriptive Family Atlas* (1886), *The Imperial Office Directory and Reference Atlas* (1889), *Cram's Big American Railway System Atlas of The World* (1891), *Cram's*

American Perception of Space

Nephew of Rufus Blanchard, George Cram published a wide variety of maps and atlases characterized by the "American look." Produced for popular reference and not as works of art, his maps were rendered in a simple style and conveyed the maximum possible information.

Big War Atlas (1898), *Cram's Bankers, Brokers and Attorneys Atlas* (1913), and *History Atlas: Story of the Great War* (1919).

The atlases accounted for the bulk of Cram's success. He continued to issue separate maps, and some advertisements list him as a globe distributor. He also occasionally made maps for other firms. Complete sets of his maps often appear in other reference books, and his atlases carry a variety of imprints other than the Cram name. Some of these were premium books such as the *United States Gypsum Company's Ideal Reference Atlas of the World* (1903) or the *Philadelphia Inquirer Pictorial Atlas* (1904). Others were probably issued for out-of-state printers with a local market in mind. The 1883 *Family Atlas*, for example, had Des Moines and Atlanta imprints. Specific editions also were prepared for special events, like the World's Columbian Exposition, and for particular states. In 1907 and 1908 Cram prepared local editions of a world atlas for at least a dozen states.

Not all of Cram's publishing ventures were successful. His roller-mounted maps for the school market never accounted for a major part of his output. In 1889 Cram became involved with a popular geography magazine, but it did not appear until 1891 and then quickly died. His business with the railroads also was limited. His *New and Correct Map of the Great Rock Island Route* (1883) was reprinted several times, but Rand, McNally's domination of the railroad map market was left unchallenged. Cram became involved with several book ventures, including *The Home Queen: World's Fair Souvenir Cook Book*, but apparently they were neither numerous nor especially successful. General maps and atlases remained his specialty.

Cram's knowledge of merchandising may have been a major factor in his publishing success. Several advertisements indicate that he continued to sell maps from his place of business as late as 1906. Complete editions of his atlases were sold as premium giveaways to various companies. Many of his publications were sold through mass-merchandising outlets: stores, mail-order firms, and newsstands. The most characteristic distribution system, however, was a corps of individual agents recruited by the company. From the early map depot days, the Cram firm had been a supply house for traveling book salesmen. *The Columbian World's Fair Atlas* carried a full-page advertisement seeking new agents: "Work and Win!" it exclaimed, "No man or woman need be unemployed...Old or Young...Everybody who can talk, can sell...No capital required. Persons who have no experience will be taught how to sell." The announcement listed various lines of merchandise that reveal the full scope of Cram's inventory: atlases, bibles, subscription books, maps, religious and juvenile charts, and stationery packets. "Every family wants something," Cram advised. The key to business success was therefore to provide a variety of popular goods at reasonable prices. The actual selling could be left to others, even large firms, like the National Map Company, which specialized in selling Cram maps. Before it purchased Cram's business in 1921, National did not publish its own maps but relied entirely on publishers like Cram for its line of products.

American Perception of Space

WORK AND WIN!

No Man or Woman need be Unemployed.

Old or Young, Everybody who can talk, can sell

CRAM'S ATLASES

BIBLES, BOOKS, CHARTS, MAPS.

STATIONERY PACKAGES, ETC.

EVERY FAMILY WANTS SOMETHING.

CRAM'S UNRIVALED ATLAS

Has a World-Wide Reputation for Cheapness and Accuracy.

Cram's Universal Atlas and Cram's Standard American Atlas are Equally Popular.

Fine Original Styles of BIBLES, In Elegant Binding, Beautiful and Cheap.
The Best and Most Easily Sold Subscription Books.
Pictorial Indexed Maps of the UNITED STATES and WORLD, Large size and scale to hang on the wall.
Finely Colored Religious and Juvenile CHARTS.
STATIONERY PACKAGES, Filled with Best Material.

Agents Wanted in Every Town and County.   NO CAPITAL REQUIRED.   Persons who have had no experience will be taught how to sell.

Address, **GEO. F. CRAM,**

415-417 Dearborn Street,   CHICAGO, ILL.

SEND FOR PARTICULARS.

Cram's most characteristic distribution system for his maps was a core of sales agents. Advertisements (here from the Columbian World's Fair Atlas*) urged everyone to apply. Success beckoned, and "no capital" was required.*

The Maps and Atlases of George F. Cram

George F. Cram is remembered primarily for his atlases. He developed the technically complex business of producing atlases, however, from his earlier experience with separate maps, which were large in size and usually folded into paper covers, often with some additional printed matter. Cram concentrated his earliest efforts on creating such maps. Cheaper to produce than atlases, the maps could be customized for special audiences. For Cram, they were a lasting financial success, and he produced separate maps throughout his career, though in declining numbers by the last decade of the nineteenth century.

Most of Cram's separate maps covered individual states, which speaks to the special nature of American federalism. Cram considered regional maps superior to state maps. In 1870 he issued a regional map, *Cram's Township and Railroad Map of North Western States*, but he soon returned to the popular state maps. The *North Western States* map was reissued in 1872, but his only other publication was an 1879 regional map of Virginia, West Virginia, Maryland, and Delaware. Even this map, however, used the list of states for its title, confirming the strong connection between states and cartography in America.

One issue raised in the mapping of the nation on a state-by-state basis was the use of different scales for maps, making it difficult to compare distances. Each state's boundaries were expected to fill out a large sheet, as if in testimony to its sovereignty. The state-by-state approach thus made it difficult for a publisher to divide up a large wall map of the nation into a series of regional maps for an atlas. Cram tried this latter approach for his first atlas in 1875, a common technique of the period. But the regional maps increased the coverage of the West, and seemed to diminish the status of the eastern states and, in general, detracted from the notion of state sovereignty. Eventually, Cram developed separate maps of each of the states specifically for atlas use. He also issued a series of larger scale state maps to be sold individually, although he did not cover every state in this format. In several instances he also pasted separately published fold-out maps into his atlases. As Cram focused more and more on the production of atlases, he issued fewer and fewer separate maps, but rapidly expanded his inventory of specific state and national plates for use in books.

The only popular American regional maps were the various route maps designed especially for railroad companies. The specific requirements of these maps as advertising pieces often led to considerable distortion: routes were straightened, distances juggled, and settlement thickened along the sponsoring line's right-of-way. The special character of route maps thus hindered their suitability for general reference use, which may explain why Cram did not produce them. The sole exception was his *New and Correct Map of the Great Rock Island Route*, issued between 1883 and 1889.

Cram's name did become associated with several other types of maps. The first imprint to bear his name was a street map of Chicago, but this was simply a version of Blanchard's publication. Cram's uncle continued to issue these local maps until his death in 1904, but beginning in 1899 Cram was listed as a joint publisher. Cram's street map of Chicago was published well into the twentieth century, but Cram himself had little to do with it. From time to time Cram also published separate maps of other American cities: New Orleans in 1892, New York in 1904 and 1908, and possibly others as well. His major contribution to mapping American cities, however, was the series of full-page maps of major urban areas he included in his atlases. *Cram's Unrivaled Family Atlas of the World* (1889), for example, contained twenty-seven maps of American cities. His huge *Universal Atlas* of 1900 doubled this urban coverage. Cram also published a variety of special purpose maps. His *New Ideal Survey Map of California, the United States and the World* (1914), a large wall map on rollers, probably was meant for school use, foreshadowing the time in the mid-twentieth century when the Cram imprint would become identified almost entirely with instructional materials.

Cram apparently issued no separate maps of a foreign country under his imprint, but he did publish several editions of world and European maps. In 1914 he rushed into print a map of the European war, following it with several other editions, including one in 1920 showing the shape of the postwar continent. Cram perceived the connection between wars and the popular demand for maps and effectively exploited it through a variety of atlases. *Cram's Big War Atlas* appeared in 1898 during the Spanish-American War; his *Complete War Atlas Covering every Possible Scene of Conflict* dealt with the Russo-Japanese conflict of 1904–5; and a whole series of atlases traced the path of World War I. Willa Cather's autobiographical novel, *One of Ours*, recalled the reaction of Nebraska farm folk to news of World War I. Her family pulled out its atlas and immediately put the events into a spatial context. Their atlas may well have been one of Cram's, who was by 1914 long established as a major publisher of popular atlases, including a complete line of pocket, business, reference, and family atlases.

The small-sized pocket atlases were a recent innovation. Most dated from the early twentieth century, and the titles indicated the nature of their appeal: *Success Handy Reference Atlas* (1902), *Cram's Quick Reference Atlas* (1906), and special pocket editions for the army and navy (1907–12). The 1904 edition of the *Success Atlas*, with 574 pages in a 9-by-15-centimeter format, shows how much was packed into these small volumes. The *Quick Reference Atlas*, in the same format, included 182 colored maps in 356 pages.

Then as now, sensing a new market was the crucial first step to business success. Cram's earliest atlas, *The New Commercial Atlas of the United States* (1875), which tried to combine a national atlas with a local business directory and met with little apparent success, may have taught Cram the importance of focusing on a particular segment of the business community. So his future production would indicate. The *Imperial Office Directory and Reference Atlas* of 1889, a joint venture with Henry S. Stebbins, filled 280 pages with a variety of data including lists of real estate agents, mortgage loan agents, banks, and hotels, along with a complete county and railroad map of every state and territory.

The next year Cram issued his first *Bankers and Brokers Railroad Atlas*, and in the first decade of the twentieth century, he published a series of shipper's guides to various states. In addition, there was the *Standard American Railway System Atlas* that claimed to show the "true location" of all railroads, towns, villages, and post offices. The reference to true location probably was an attempt to correct the distortions often found on the railroad route maps. By 1910 this huge reference volume comprised 636 pages with 150 colored maps.

Several of Cram's later reference atlases were even larger. The new census editions of Cram's *Atlas of the World: Ancient and Modern* (1901, 1902, 1903) extended more than 836 pages and included a geographical encyclopedia. The book's introduction claimed that it was the first comprehensive atlas produced entirely in the United States. In addition, Cram announced that the services of his firm's Bureau of Geographical and Commercial Information would be available to purchasers of the atlas.

Cram's general reference atlases appeared under a variety of titles. The only atlas that retained a similar title over the years was Cram's most enduring success: the *Unrivaled Family Atlas*, which first appeared in the early 1880s and reached its 64th edition in 1952. Its contents changed and grew with each edition. The earliest books contained about a 100 pages, while later editions were much larger. Cram always emphasized the size and scope of his family atlases, which had to be more complete than the smaller premium atlases he issued after 1892.

In 1907 and 1908 Cram experimented with special state editions of a general atlas, volumes which varied in length but had similar titles, such as *Cram's Superior Reference Atlas of Kansas and the World*. Individual volumes appeared for California, Nevada, Iowa, Illinois, Michigan, Missouri, Nebraska, the Dakotas, Oklahoma, Wisconsin, and Minnesota, each with special maps and charts on the geography and economy of the state, historical and pictorial sections, and a large fold-out map of the state pasted on the back cover. A detailed index to this map together with data on railroads and shipping services made up the last section of the atlas. None of the "superior state" atlases reached a second edition, although almost a decade later a similar approach was used for special publications on New York and Pennsylvania.

Maps and the American Perception of Space

Nineteenth century Americans, and other people in the western world, generally viewed the larger world around them with keen interest and some anxiety. Maps measured this world. The

---

Opposite: Cram's Superior Reference Atlas *was one of many such volumes produced by Cram in Chicago and sold nationwide. Also a producer of flat maps, Cram was known for utilitarian and reasonably priced work that catered to a fast-growing middle-class market.*

American Perception of Space

presence of an atlas in most middle-class homes and in many working-class ones bespoke this new sensibility. Somewhere in a bookcase, on a shelf, or gracing a table, the treasured family atlas held within its ornate covers a compendium of information about the wider world. The family could gather around this treasure for instruction and edification. Indeed, some of Cram's atlases incorporated just this domestic scene into the design of their title pages. The 1889 *Unrivaled Atlas*'s title page featured a family around a table looking at an atlas. An atmosphere of quiet earnestness in the quest of culture pervades the scene. The bookshelves in the background and the paintings on the wall amplify the point: people commonly learned within the family, and to pursue this mission effectively each household needed an atlas. The mother is seated at the table, imparting knowledge to her son. The image of classical ruins on the title page implies that the atlas extended an important function in the transmission of culture. The ship in the harbor indicates exploration and commerce reaching the most distant places on the globe. Everything evokes an atmosphere of confident adventure. By way of the atlas, thousands of families traveled in their imaginations to distant exotic places, thus the volcano above and the waterfall below. The Indian encampment in the background gave the volume an American flavor.

An atlas is, above all, a way to organize knowledge about the world, and the way in which this information is packaged suggests a particular cultural orientation. Atlases organize geography for their readers who find in them a particular world view. That view, if the atlas is commercially successful, will probably confirm their notions of what the world is like. The late nineteenth-century atlases portrayed an orderly though changing world in which all the pieces fit together as progress marched forward.

Once cartography is put in its cultural context, a number of questions arise about the cartographer. In this case, what did Cram want to communicate by his maps and atlases? Did he have a conscious idea of the language his maps spoke? Or did the market alone govern his publications? The success of his atlases indicates that they struck a responsive chord among the American people; thus his maps are excellent sources for

Maps organized space, sometimes in unusual ways. Cram's Atlas of Illinois and the World *(1906) offered readers this view of the relative sizes of the United States and Europe.*

discovering the American perception of space. Although Cram's maps appealed to the mass audience, they differed—sometimes markedly so—from the maps produced by other firms. They carry what scholars have characterized as "the American look." Because they were produced for popular reference rather than as works of art, the design was usually very simple. Ornamentation gave way to information as the maps filled up with as many place names as possible. Rather than focusing on a few elements to make the maps strongly expressive, Cram tried to be as complete as possible by leaving nothing out and including something for everybody. The decision to mark his maps with so many names, giving them a dense visual quality, indicates that Cram saw the location of places as their primary function. Almost all of his maps were political in

American Perception of Space

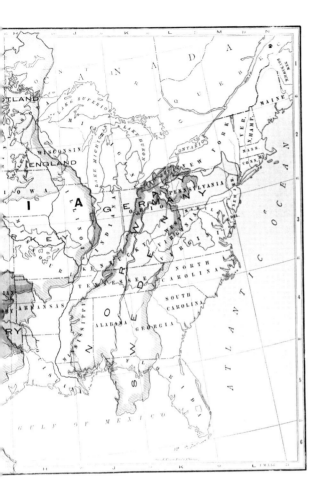

nature, with national, state, and county boundaries marked in prominent lines and colors. A mobile population demanded maps to plan where it was going, to help find its way there, and to orient itself once it arrived. Thus, Cram focused on maps both of the new western states and of the country's growing cities and towns. The railroads, which linked them together both visually and literally, always were featured prominently.

The American perception of space as suggested by Cram's work entailed a view of origins as well as prospects. The United States was, after all, a nation of immigrants, and a large portion of Cram's market was born abroad or had roots in another country. Unhesitatingly, Cram tapped this interest. Likewise, he capitalized on the establishment of American colonies overseas in the 1890s, emphasizing these areas in his books. Special attention also was given to the Holy Land, and biblical maps were always an important part of the larger publications. Churches as well as homes and schools were a growing market for his atlases.

Cram's focus remained, however, on the western world and particularly the American states. Thus, his work reflected a resurgent nationalism at the same time that it highlighted the importance of the separate states. Like European nations and indeed whole continents depicted in maps, each state received a full page in Cram's atlases. Cram also realized that maps linked to history were even more compelling, and his atlases regularly featured lists of significant dates in the annals of America, along with portraits of the presidents and other historical material. Cram's maps of the world on the Mercator projection placed Washington, D.C., in the center. Longitude was then computed from Greenwich on the top of the map and from the national capital on the bottom. In the eyes of most of Cram's readers, America stood at the center, and Cram's atlases fit comfortably into this perception of America. For a people who craved information and liked to keep up-to-date, they were fact-filled and affordable. New census editions poured forth, aimed at the same market which had bought old ones. Cram regularly advertised them as "unrivalled"—hyperbole he and his readers probably believed. Unrivalled or not, they were irresistible and useful tools for thousands of Americans bent on bettering themselves through education. For Cram, the publisher, they were the means to success. Whether he ever thought of himself as a shaper of American cultural consciousness, we do not know. But like all mapmakers' works, his assumed a life of their own, molding as well as reflecting a distinctive American perception of space.

For Further Reading

THE MASS APPEAL of maps in the nineteenth century is, as yet, a largely unexplored field. David Woodward's *The All-American Map* (Chicago: The University of Chicago Press, 1977) provides a splendid introduction to the wax engraving process and the mass production of maps. Little documentary evidence on Cram or his company has been located. The current Cram firm in Indianapolis does not have any business records from the early period, and any personal papers that Cram may have kept also have disappeared. Thus the primary sources of evidence are the maps and atlases themselves, which are scattered across the country in numerous depositories.

45

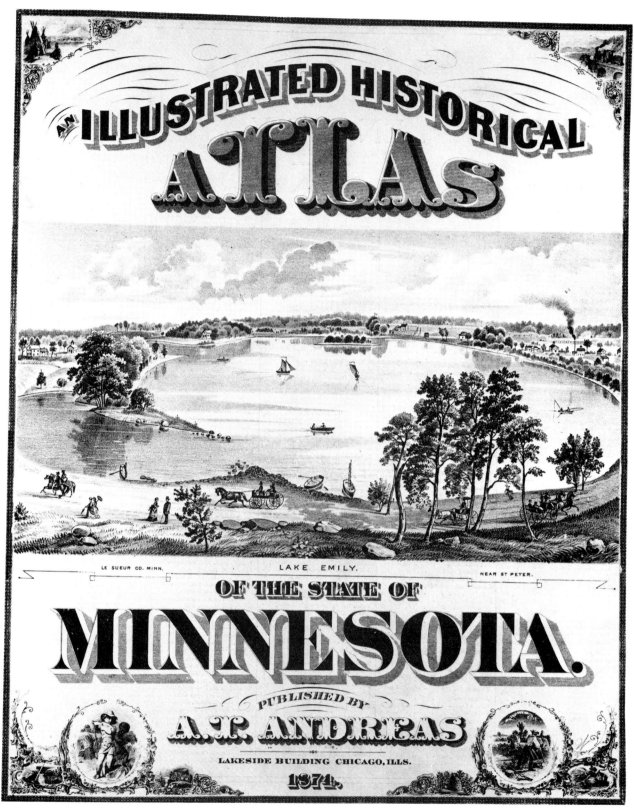

Title page from Andreas's Minnesota atlas of 1874. Andreas applied his highly successful formula for county atlases to create this publication, the first of his state atlas series.

Maps for the Masses: Alfred T. Andreas and the Midwestern County Atlas Trade

By Michael P. Conzen

The story of Alfred T. Andreas provides an intriguing glimpse of American mapmaking and the ways in which advances in printing, marketing, and business practices shaped the role of maps in the lives of ordinary people.

A VERITABLE ARMY of "canvassers" or traveling salesmen crisscrossed rural America in the nineteenth century, leaving in their wake not only the goods they peddled but also a rich folklore of traveling salesman stories and the not infrequent aftertaste of swindle. Along with the trinkets and ribbons, bibles, patent medicines, encyclopedias, and lightning rods, there was another item that took salesmen out on their rural rounds—the elaborately illustrated county atlas, featuring maps showing individual landownership and lithographic views of property, which were sold by advance subscription. Alfred T. Andreas was a major figure among publishers selling county atlases in the early 1870s, and his activities were to draw animated comment. "It was a scheme that, in five years, took from every county in the northern States numbering 10,000 inhabitants and over, not less than $10,000, and in some cases as high as $45,000," complained Bates Harrington in a muckraking exposé of county map salesmen published in Chicago in 1879. The implied charge of wanton profiteering by atlas makers, however, reflects neither the full financial story nor concedes the creative imagination and entrepreneurial risks that underlay this special kind of business.

Chicagoans' familiarity with Andreas stems largely from his monumental three-volume *History of Chicago* (1882–1886), but this comprised only the finale to an extraordinary publishing career that produced county atlases, state atlases, and later county, state, and city histories on a grand scale. It was a tempestuous career, building small fortunes only to lose them in later bankruptcies, but in just a few short years Andreas revolutionized the art of county landownership mapmaking and helped establish Chicago as the nation's leading center of subscription publishing and atlas production. This alone warrants his place as a pioneer in the history of American cartography. His enigmatic character, forceful leadership, and peculiar strokes of fortune, however, provide a particularly enlightening glimpse of American mapmaking after the Civil War and illustrate the manner in which advances in printing technology, marketing, and business organization affected mapmakers and the role of maps in the lives of ordinary people.

Historically, maps have served to orient people to the physical arrangement of the world around them. The variety of things maps have shown and how they have shown them is almost endless, but until comparatively modern times maps were intended for well-educated people and those with special occupations—kings, navigators, military commanders, and others in authority. Governments have been crucial in stimulating advances in cartography, and this was true in the United States as its territory and economy grew during the nineteenth century. But the demands of exploration, military preparedness, and initial land survey left many spheres of mapmaking to private initiative and market demand. The development of printing by lithography enabled mapmakers to offer general reference and regional maps to the public at unprecedentedly low prices, and the county landownership map emerged at mid-century as a major new form of popular map. It was attractive to real estate dealers, local officials, and

Michael P. Conzen is associate professor of geography at the University of Chicago.

residents interested in their surroundings and could be made profitable by advance subscription sales. The true key to its mass appeal, however, was its transformation into a vehicle for the cultural expression of pioneer pride, material accomplishment, and civic self-congratulation—a winning formula that contributed to and benefited from a rising historical consciousness during the nation's centennial era.

Among the small, scattered group of men responsible for nudging the county landownership map in this crucial direction, none surpassed Andreas in creativity and missionary zeal. In five frantic years he made county atlases big business, put their production on a factory basis, and developed subscription marketing into a high art form, which set a standard for the industry.

"A Cosmopolitan Yankee"

Alfred Theodore Andreas was born on May 29, 1839, and raised in small-town upstate New York, first in Amity, and then in Chester, Orange County. Though his father, William H. Andreas, ultimately abandoned small-town storekeeping for a successful business career in New York City, young Alfred struck out for the West at eighteen. He spent three years teaching school in Dubuque, Iowa and later settled in the southern Illinois town of Sparta. There he enlisted in the Twelfth Illinois Infantry as a private on July 21, 1861, giving his occupation as "merchant," and within two months was assigned duty in the quartermaster's department. Promotions came quickly: to commissary sergeant in May 1862, by commission from Governor Yates; 1st lieutenant and regimental quartermaster on January 1, 1863; and acting commissary of subsistence of the 2nd Division, 16th Army Corps, in May 1864. After participating in the Atlanta Campaign and Sherman's "March to the Sea," he was mustered out at Goldsboro, North Carolina, on April 1, 1865, and breveted a captain at the age of twenty-five. His army experience introduced him to organizing complex movement of supplies and dealing with large numbers of individuals. Andreas thrived under these pressured conditions and was to apply his experience in his civilian life. The "cosmopolitan Yankee," as one chronicle described him, was now in need of a permanent career.

Eight weeks after his discharge, Andreas married Sophia A. Lyter of Davenport, Iowa, in the Davenport Christian Chapel. Then, with a nineteen-year-old bride, he cast around for a living. By December 1865 he hit upon the idea of establishing a winter skating rink on bluffs at the top of Brady Street Hill in North Davenport overlooking the Mississippi River. It was an ill-starred beginning. First, the wells he sank for water failed to provide. Then, after he had river water carted more than a mile up the hill, three days of hauling came to nought as the newly graded ground in the rink absorbed the water with ease, returning it to the Mississippi at no extra charge. He gave up on December 22, and Christmas that year must have been subdued.

Andreas remained in Davenport professionally uncommitted until Louis H. Everts, a former army associate, persuaded him in 1867 to canvass for the county map publishing firm of Thompson and Everts of Geneva, Illinois. County mapping had been flourishing in the East since 1848, and some eastern surveyors like Henry Walling, Frederick Beers, Samuel Geil, and other associates of Philadelphia mapmaker Robert Pearsall Smith were by then venturing into the Midwest. Many western maps were made by local men, however, notably Moses H. Thompson, who, beginning in the late 1850s, had mapped several Iowa and Illinois counties with his brother, Thomas H. Thompson. After the Civil War Thomas formed a partnership with fellow townsman and army associate Everts—both had served in Andreas's army division—and were busy making county maps in Jones and Jackson counties, Iowa, when Andreas joined the firm. These maps portrayed rural landownership, local topography, and selected artists' views of individual land holdings and were sold by advance subscription, hence the need for canvassers. Andreas's aptitude for map peddling was at first uncertain, but his personable manner with others soon made him shine.

While selling map subscriptions in Delaware or Linn counties, Iowa in 1869, a great idea struck him. If the county map were broken up to show township segments on individual pages in an atlas, it would be much less bulky, and would permit unlimited farm views and other material to be printed on intervening pages. An atlas would be easier to consult and store, and the expandable space for illustrations in an atlas promised increased revenue through the extra fees charged for their inclusion. It also

Creator of atlases designed to appeal to the broadest possible audience, Alfred T. Andreas democratized a genre of mapping once confined to the professional and cultivated classes.

offered the prospect of covering successful old territory again with a new product.

The first true county atlas is believed to be that of Berks County, Pennsylvania, published by Henry F. Bridgens in 1861, but the circumstances surrounding the origin, acceptance, and diffusion of the county atlas genre are anything but clear. Andreas may have seen one of the eastern atlases by the Beers family on his travels; he had family connections in New York State where twenty-three county atlases had appeared in 1869. Perhaps he discovered one of the three unusual atlases by Lyman G. Bennett produced in Minnesota in 1867–68. More likely, he might have come across Nehemiah Matson's 1867 history-cum-atlas of Bureau County, Illinois, only two counties distant from his own canvassing activities, or the New Coles County, Illinois, atlas (1869) by Warner and Higgins, easterners who had worked their way west. Or, it may be that Andreas did indeed stumble upon the notion independently, as some contemporaries have claimed.

Whatever the case, the atlas idea was attractive. Though skeptical, his employers were impressed enough with Andreas's determination to ask him to stay with the firm until an atlas experiment in one county had been tried. According to one source, Henry County, Iowa provided the test case.

Apparently the experiment was a success, for Andreas promptly withdrew from the firm to make his own arrangements. With an exciting business prospect now firmly in view, Alfred Andreas reached back to his wife's family in Davenport for financial support. He formed a partnership with his brother-in-law, John M. Lyter, and also with F. H. Griggs, one of Lyter's partners in Iowa's largest printing firm, Luse and Griggs. Between 1862 and 1869 the firm had parlayed its $8,000 capital into $50,000. Reorganization with new partners in 1869 increased the capitalization of Grigg's printing operation to $80,000, rendering him quite capable of acting as a sleeping partner in the new firm of Andreas, Lyter and Company, formed the following year. Griggs naturally provided the printing facilities.

Thus organized, Andreas plunged into county atlas production and mounted a campaign in Knox County, Illinois, a well-settled and rich farming region in the Military Tract. Using a doctored copy of the Henry County, Iowa atlas as his sales dummy, he expanded the concept of a county landownership atlas with new features. To the familiar mix of township landownership maps with select business directories, state and county maps, village and city plats, brief historical material on the county, and some lithographic views, he added a copious subscriber's list giving each patron's name, residence, business, nativity, year of arrival in county, and former residence. In addition, there was a digest of census statistics for the county, and a page at the back for recording transfers of real estate. When this last item proved no special draw it was dropped quickly. Attention to detail was notable: besides the standard property divisions, houses, rivers, roads, churches and schools, the maps included elements such as mills, blacksmith shops, quarries, lime-kilns, woodland and swamp, and coal outcrops. The subscribers' list added a major new dimension to the appeal of the county atlas. Most important of all, Andreas made great efforts to solicit orders for lithographic views of farms and businesses. The traditional county maps had generally contained a few views of major institutions, such as the courthouse, and some residences of local grandees. Andreas's touch was to democratize this aspect by encouraging humbler citizens to be represented, and it proved a Midas touch.

The Knox County atlas was a major success, with sales to 1,225 residents, and Andreas switched operations without pause to Fulton County, immediately to the south. Once assembled, it was essential to keep his survey and sales forces together and functioning. The small group of field surveyors—anywhere from two to five in number—began by making copies of the government land offices' township plat maps which would provide a map base, survey grid lines, and natural features such as river courses, bluffs, and marshes. A second stage consisted of drawing from the land records or tax lists in the local county courthouses landownership information, which was later recorded on the maps by placing owners' names over the appropriate land parcels on the map. There they would also copy village and city plans from the plat records. A third stage involved riding over all parts of the county in order to sketch in on the map from direct observation the cultural features such as roads, houses, railroads, town sites, schools, mills, and woodland cover.

Meanwhile, the sales force scoured the countryside and the towns securing subscriptions. Local officials, businessmen, lawyers, doctors, and newspaper editors were approached early for endorsements of the project, which counted heavily in persuading others to subscribe. Appeals were made to vanity and patriotism. The atlases recorded the establishment of individual families on the land, expressed their pride of ownership, and asserted their role in the pioneer history of the region. They also mirrored a resident's standing in the community and advertised his assets, particularly his property holdings. Businessmen and professionals could have their services lauded with dignity, and the district at large could put on its best appearance, boast its "improvements," and advertise to the world its civilized state.

The wandering canvassers toured county settlements in careful sequence: first the "lightning man" who made the key contacts, then the subscription agent, then the "viewer" who made the lithographic sketches, and finally (in later years when two more elements had been added to atlases) the biography editor and the portrait canvasser. Then as now, it was not so much the basic product but rather the optional extras that really turned the profit. The base subscription generally cost $9, for which a subscriber received a copy of the atlas with his name engraved on the map and listed in the "subscriber's list" or "patron's directory." Views, portraits, biographies, and business notices cost extra. The four-inch "view," the smallest available, cost $28, while a full-page view cost $145. Portraits came at $250 or $100 each, and biographical material was assessed at two-and-a-half cents per word. The illustrations contributed over one-third of the revenue for an Andreas atlas.

When all the selling, sketching, and biography-taking had been done, all the mapping and checking had been completed, and all the collation and arrangement of material was ready—only then was the print run decided. Then came several weeks for production, shipping, delivery, and collection of payments. The entire process took from three to six months, depending on the size of the work force and external factors. Total subscriptions set the size of the edition, and therefore the overall costs. The revenue depended upon collections, and economic conditions at time of delivery often affected the ultimate profitability of an atlas.

Andreas found county atlases unquestionably profitable. He made enough money to invest in an increasing number of atlases per year. His geographical strategy was concentrated, as he extended his operations to neighboring counties, in order to give his work force a coherent program with minimum travel. Within three years he produced eleven atlases, all the counties in western Illinois arranged in a contiguous, snake-like progression from Knox to Jersey counties, and then starting a northern swing west from Tazewell County.

With his atlas production systematized Andreas was now ready for bigger things. Without breaking stride he dissolved the company in 1873, bought out his partners, and moved his operations to Chicago. By May net profits had amounted to $47,000, and even after purchasing his partners' original shares, giving them 10 percent interest, a $20,000 cash bonus, and covering

An illustration from Andreas's Tazewell County atlas (1873) showing an idealized view of an Illinois farmstead and, inset, a map of the subscriber's (J. Merriam) holdings.

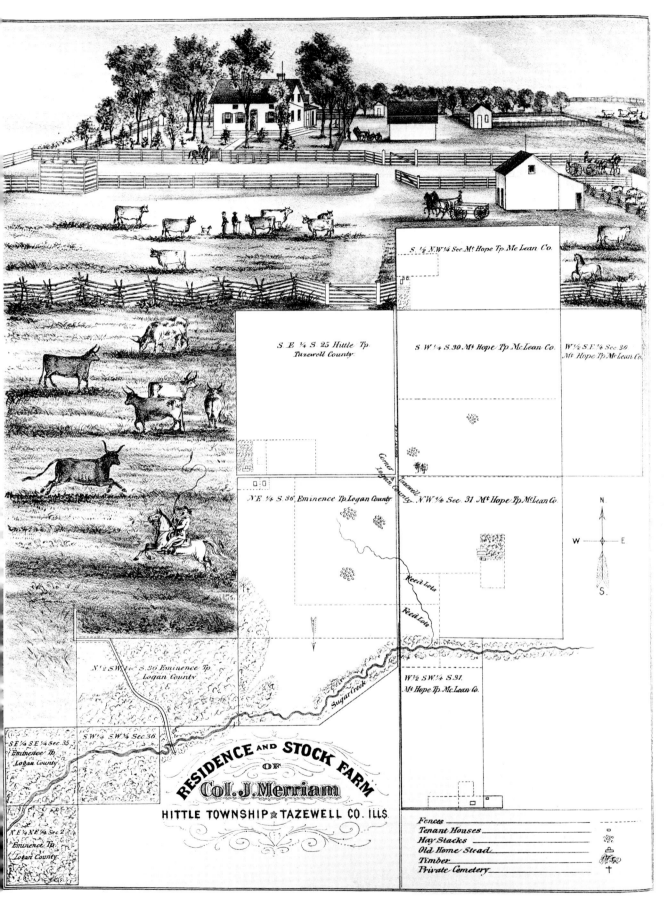

Maps for the Masses

COMMON SCENE IN DELIVERING ATLASES.
SUBSCRIBER'S WIFE.—"Fifteen dollars for that 'ere big book, and too poor to ever get a cook-stove, or fix the house up!"

County atlas canvassers selling subscriptions commonly appealed to the farmer rather than his wife, whose household priorities did not allow such a luxury. Thus, often it was not until the atlas was delivered, as shown here, that a subscriber's wife learned of her husband's investment.

salaries and expenses, Andreas was positioned to begin in Chicago with a capital of $20,000.

The desirability of relocating in the region's major metropolis stemmed from events in the county map and atlas publishing field as a whole. Since 1864 the Beers family and their associates had been producing county atlases on a moderate scale in New York and Pennsylvania, and a few in Ohio. In the Midwest a few isolated atlases had been produced before 1870, but the switch from maps to atlases made in that year by Warner and Higgins, Andreas, and Thompson and Everts galvanized map publishers throughout the region. From then on most county mapping was in atlas form, and maps appeared only for small counties or in frontier zones with low population densities. Atlases cost almost four times as much as maps to produce, but if sales could be doubled or local wealth result in profuse illustrations, the higher unit price of atlases would return profits three times larger for the effort. But the uncertainties of catering to these highly segmented, local markets with individually tailored pro-

ducts quickly took their toll of publishers. Those based in large cities with ready access to experienced map printers, labor pools for agents, and extensive short-term credit sources were most likely to be successful.

Thus the surveyor gave ground to the versatile businessman, and many map companies prospered because their scale of operations and business connections gave them durability as well as competitiveness. Andreas had none of the skills of the surveyor-engineer, draftsman, or printer's engraver, but his talent for organization and his business sense pointed him firmly in the direction of the big metropolis, and his ambition towards larger plans. Besides, a year earlier Everts had bought out his partner Moses H. Thompson, and, leaving Geneva with two former employees, Oliver L. Baskin and David J. Stewart, had preceded Andreas to Chicago to continue the county atlas trade from there. Already established in the city since 1869 was the third major midwestern county mapping firm of Warner and Higgins (later Warner and Beers), who suffered a great loss in the Chicago Fire of 1871. Out of the ashes of that conflagration emerged the Lakeside Building, an imposing six-story edifice at the corner of Clark and Adams streets that was to symbolize the dynamic reorganization of western atlas publishing and steal much of the cartographic limelight for Chicago that Philadelphia had hitherto enjoyed. The Lakeside Building was explicitly designed to house firms representing all the various manufacturing phases in printing and atlas production. It was a revolutionary concept, characteristic of the restless energy and spirit of Chicago in the wake of the fire, and Andreas was drawn to it.

Going for the Big Time: The Chicago Atlas Years
Chicago's map trade matured quickly after the Civil War. The days in the 1850s when Rufus Blanchard sold eastern-made maps wholesale and retail and published a few of his own on the side were long gone. By the early 1870s, dealers and publishers were largely independent, some printers had specialized as map printers, and many publishers were concentrating on particular types of maps. Andreas felt in his element. "It did not require a prophet to foresee," he wrote in his Minnesota atlas of 1874, "that Chicago was destined to be the great western center of literature as well as of commerce, and the Lakeside Company determined to provide the amplest facilities for the publication of atlases, maps, gazetteers, books, magazines, newspapers, etc., etc., to supply the ever increasing demands of the people." The field surveys and subscription lists coming in for his county atlases were collated, edited, and prepared for publication in his offices on the Lakeside Building's third floor. Atlas maps were then sent up to the fourth floor to be engraved and lithographed by Charles Shober's Chicago Lithographing Company, and returned to the third floor for coloring by Warner and Beers. The text was sent up to the sixth floor under the mansard roof for printing by the Lakeside Publishing and Printing Company, and leather atlas covers were sent to the basement for gold tooling by A. H. Reeves. Finally, the printed pages were bound by A. J. Cox and Company on the fifth floor. Such concentration under one roof produced publications "executed with an economy, promptness and beauty otherwise unattainable."

So installed, Andreas set about realizing a dream that must have been maturing in his mind for some time. If county atlases represented greater profits than county maps, could not an additional increase in scale of publication augment returns even further? His thoughts turned to state atlases. As before there was some precedent to follow, and he could see great room for improvement from the business perspective.

Henry Walling had turned to publishing state atlases in 1868, thereby outflanking his many competitors in the county map field. His atlases of Ohio (1868), Illinois (1870), and Michigan (1873) as well as selected eastern states represented significant bench marks in the topographical mapping of these regions. Reflecting Walling's training and interests as a professional surveyor, they consisted almost entirely of county "outline" maps showing survey lines, administrative boundaries, roads, and so forth, and were largely unillustrated. Andreas saw a broader purpose in state atlases: they would be modified extensions of his successful county atlases, oriented to the direct interest of the ordinary farming and business population. This meant voluminous pictorial views of farms, businesses, local institutions, portraits, and the all-important subscriber's lists.

Andreas's publishing strategy changed dramatically after coming to Chicago. While issuing a few more atlases of counties adjoining those in his previous collection (Peoria and Hancock, Illinois; Des Moines, Louisa, and Lee, Iowa), he began producing atlases of widely separated contiguous pairs of midwestern counties in addition to launching his first state atlas. This scattering suggests a focusing of effort on carefully chosen counties that were "safe bets" to make money while planning and investment proceeded with the state atlas idea. Besides, competing firms were busy staking out regional territories to exploit as Andreas had done in western Illinois and eastern Iowa. Everts "mopped up" in northern Illinois and southern Wisconsin, and then swung through Michigan and Ohio while moving his headquarters permanently to Philadelphia. Warner and Beers persevered with what remained of the profitable counties in Illinois. Higgins and Belden plied Indiana, and the Edwards Brothers worked their way through northern Missouri.

These publishers lost no time in converting to

the county atlas format for their landownership maps, but Andreas had the inspiration to stay a step ahead of the competition. He boldly chose Minnesota for his first state atlas effort. Precisely because Minnesota was still largely a pioneer realm in the 1870s and had barely seen any county map or atlas activity, he saw great possibilities in its "unatlased" condition. A state atlas would of course contain maps of whole counties, and because their scale precluded showing all landowners in a district, Andreas included on the maps only the house locations, names, and acreages of actual subscribers.

Andreas hired his former boss, Thomas H. Thompson, to superintend the field aspects of the operation and established a local office in St. Paul. A work force that eventually included 108 people labored to produce this novel atlas in less than two years. Essentially, Andreas employed the methods he had perfected with his county atlases, with refinements and extra flourishes. Fieldwork had to be conducted on a gigantic scale and was not easy in districts west and north of Minneapolis where roads and railroads were scarce and population thin. Consequently, Andreas combined several counties on one map, and the whole of northern Minnesota north of Douglas County appeared on one double-page map.

The experience of the field survey team was duplicated by the canvassing team. Higher per capita sales were recorded in settled than in frontier counties. The extra flourish Andreas brought to the Minnesota state atlas involved greater use of newspapers in the sales campaign than before. Since a much larger territory was involved, stimulation of news coverage about the atlas preparations was essential to raise awareness and interest. After an experimental canvass in selected counties proved highly promising, Harrington records that "a man was then sent to secure the newspapers. He made a contract with every newspaper in the State, at very low rates, to print all articles sent them by A. T. Andreas, at so much per line. Simultaneously, every sheet in Minnesota loaded its columns, extolling to the skies the grandest enterprise of the day and age. People read, became enthusiastic, and longed for the day when they could gaze upon such a production." These articles often featured endorsements by public figures.

The general success of the enterprise can be gauged by the 12,000 subscriptions secured, and the 350 pictorial views, 340 portraits, and 219 biographies commissioned for the volume. The atlas was conservatively expected to gross about $250,000. Costs have been calculated indirectly at $200,000, leaving a $50,000 margin of profit to materialize when delivery and collections were made. However, slight delays in the production schedule together with the prolonged process of delivery coincided with the deepening agricultural depression following the financial panic of 1873. By late March 1875 at least 1,000 Minnesota atlases had yet to be delivered and with remaining collections getting hard to make, Andreas was already discounting their probable value by 25 percent.

Though the Minnesota atlas turned out a mixed financial blessing for Andreas, as a publishing accomplishment it was a triumph. In no previous instance had citizens of any state been presented with a nearly 400-page atlas in which the geography, history, and official statistics of their domain were so amply displayed, and so richly larded with the names, farm locations, migration history, pictorial views of property, portraits, and biographies of their fellow citizens, ordinary and not so ordinary. Andreas's sales campaign elicited atlas subscriptions from one out of every seven of Minnesota's 85,000 households, with even higher "market penetration" in the well-settled southern counties of the state—a remarkably broad appeal given the contemporary value of the $15 basic subscription price.

The purely cartographic quality of the atlas was mixed. The lithographed maps by Charles Shober were clearly professional but lacked the elegance and fine detail of Mendel's work on Bennett's atlases of Winona, Ramsey, and Olmsted counties seven years earlier. Still, the county maps were printed by chromolithography, which added a certain opulence. The artists' views of subscribers' urban and rural residences, businesses, and institutions, and the bird's-eye views of towns were executed with competent but unexceptional skill. Occasional infelicities of pagination and space allocation gave the atlas an unmistakable underpinning of utility: space was not to be wasted, even if the "packing" showed. Nevertheless, the sheer massing of the varied content and the cumulative visual impact of the carto-

Maps for the Masses

After the fire, many of the specialized services required for commercial mapmaking—engraving, lithography, binding, printing, coloring, tooling—came together in the Lakeside Building at Clark and Adams streets. Andreas, located on the third floor, relied on these facilities to launch his map publishing ventures.

graphic and artistic iconography of so vast a territory within the covers of a single volume was stunning.

In part to cope with the division of interest and resources between state and county atlases, Andreas went into partnership with Oliver Baskin (an old associate from earlier Thompson and Everts days) so the latter could concentrate on producing some Indiana and Ohio atlases. Only once did Andreas stray as far as the East Coast, and that was to co-publish in 1875 (with Baskin and G. S. Burr) an atlas of Orange County, New York, surveyed by Frederick W. Beers. Senti-

ment must have been a factor in persuading him to contribute to an atlas of the county of his birth. Essentially a Beers product, it had the delicacy and aesthetic appeal of an eastern atlas, but lacked features (such as woodland cover) that Andreas had enshrined in his Midwest productions.

The new genre of western state atlases required extensive planning and logistical coordination. By late 1874, with the Minnesota atlas nearly completed, Andreas turned to an Iowa atlas and laid the groundwork for a third, of Indiana. With projected costs for each atlas of about a quarter of a million dollars, the publisher needed

Maps for the Masses

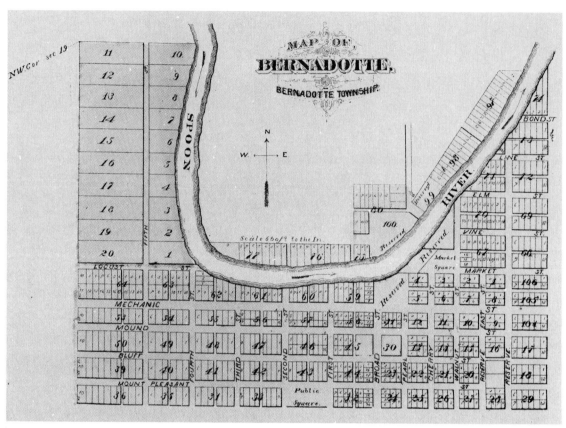

Above: *Map of Bernadotte, Illinois from Andreas's* Atlas Map of Fulton County, Illinois *(1871).* Below: *For an additional fee, Andreas offered subscribers the chance to embellish the view of their property with portraits, as shown in this example of the Burgetts of Lewistown, Illinois.*

access to unprecedented amounts of capital. A confidential credit report on Andreas the following year concluded that he "appears to be a sanguine man and inclined to lay out more business than he can control with his own means." Returns from the Minnesota atlas were to be plowed into the Iowa atlas, and so forward. To the extent that he needed additional cash, he came to depend on an old Iowa associate, B. F. Allen, a millionaire Des Moines banker with extensive business interests from Nebraska to New York City, who received a proportion of the profits. The relationship suffered an unexpected and ugly reversal in early 1875. As a favor, Andreas had endorsed a sum of $30,000 for Allen's temporary use elsewhere. Then Allen went spectacularly bankrupt. Well respected as a capitalist in Des Moines, he had like Andreas succumbed to the lure of Chicago, and in 1874 bought the Cook County National Bank, only to find it in unstable condition following the panic of the previous year. When the bank failed, Allen was ruined. Andreas lost his long-term backer as well as his own $30,000 in one fell swoop.

Thus by March 1875 a cash-flow crisis of major proportions loomed. Andreas owed more than $100,000 to creditors; while $80,000 worth of receipts expected from delivery of the remaining atlases and delayed payments could be anticipated, it might take a year to materialize. He already had invested $45,000 in the Iowa atlas, with much more needed to complete the project, and his county atlas projects with Baskin had run up debts of $45,000, which he hoped to cover with expected profits from the Orange County atlas. His solution to this dilemma was to form a stock company that would bring in fresh capital as well as secure his creditors by making them stockholders.

It is a measure of his business reputation that he succeeded. On March 24, 1875, the Andreas Atlas Company was organized with $125,000 capital and a board of directors composed of several major Chicago businessmen (including two generals) and some of his largest creditors. This event betokened both confidence in Andreas on the part of his associates and creditors, as well as the realization that a completed Iowa atlas was far more likely to return their debts due than any forced bankruptcy.

Thus revamped, Andreas's firm was able to climb out of the pit precipitately deepened by Allen's failure and proceed vigorously with the Iowa atlas. The final product was identical in conception to the Minnesota atlas, but even grander in scale—nearly 600 pages. It sold more than 20,000 subscriptions, did exceptionally well in generating views and portraits, and was delivered in October 1875. What profits emerged from this are unclear, but projected revenues around $400,000 should have more than covered costs estimated at around $300,000. It is highly unlikely, however, that collections were anything near perfect in most parts of the state.

Meanwhile, progress on the Indiana atlas at that time is not altogether clear. It appears that Andreas intended to complete this third state atlas, after which the Andreas Atlas Company would be reorganized again. It was, in fact, dissolved in December 1876. While the title page of the Indiana atlas was unmistakably an Andreas design, his name is missing from it, suggesting an eleventh-hour withdrawal as publisher-of-record. Mounting problems ranging from rainy weather to crop failure and financial drought placed the atlas's success in jeopardy. His preoccupations in Iowa did not help. Baskin's role in directing the Indiana atlas steadily grew, though he was constantly short of working capital, being no capitalist himself. Ultimately, he formed an atlas company under the title Baskin, Forster and Company, which included Frank Forster and Charles Ogelsby, both members of the paper firm of Oglesby, Barnitz and Company. The atlas generated a disappointing 12,000 subscriptions and the level of collections upon delivery was so depressing that the venture was a financial failure. Apart from a later and unremarkable atlas of the two Dakotas (1884), this ended his atlas career. Andreas, meanwhile, suffered reverses in his personal finances as well, having, among other things, lost $6,000 in some California enterprises by July 1876. By Christmas the Andreas Atlas Company closed up, having paid creditors 60 cents on the dollar. Trade circles considered Andreas "to be used up finanically."

It was a sad conclusion to a bold experiment. In five years Andreas had conceived and produced twenty-three county atlases illustrated in a lavish style, developed to a degree anticipated by none, and imitated by few, and he had created three state atlases on a grand scale seen neither before nor since. These atlases gained wide popularity, but the fad played itself out quickly in some districts, and the prolonged agricultural depression hastened the eclipse of the genre's novelty. In reaching for statewide coverage, Andreas assumed risks the market would barely support. He was a victim not of flawed design, but of wayward circumstance. Atlases of Illinois (1876) by Warner and Beers and Wisconsin (1878) by Snyder, Van Vechten were similarly pecuniary failures. Andreas built up great profits from his early county atlases only to see them evaporate in the cauldron of his ambition.

After the Centennial: "Mug-book" Mania
Some in Chicago were quick to write Andreas off, but he was, in fact, far from finished. One of the tenets of subscription publishing is that the books offered for sale should have universal appeal and be timely, or timeless. The very nature of subscription canvassing results in large regions being "worked over" so that similar ventures are unlikely to succeed in areas already covered until time has elapsed to allow demand to recover. Having a different product helped, of course, and Andreas's atlases, rich in illustrations and appeals to personal vanity, had been particularly successful in reworking old county map territory. Then, just when the wreck of his atlas schemes seemed to finish him as a publisher, President Grant's Centennial Proclamation focused national attention and approbation on the suitability of preparing local histories. Andreas did his mental arithmetic again, and moved into the largely new field of commercial county histories. He may well have been the first entrepreneur to grasp the potential of a concerted, large-scale attack on this "text-only" market by a publisher experienced in standardized but lavish book production, targeted to local communities and sold by efficient subscription methods.

Publicly embarrassed by his financial reverses, he induced an acquaintance, Harry F. Kett, treasurer of the Calkins Champion Washing Company, to join him in forming a new firm to publish county histories, to be known as H. F. Kett and Company. This gave him working capital and a respectable business front. The county histories would be smaller and less bulky than the atlases, half the price ($7.50), contain some historical narrative and only a simple county map, and emphasize biographical material concerning the subscribers. By abandoning the laborious and skilled preparation of large-scale landownership maps, Andreas saved considerable initial investment in expensive cartography. A page of type cost a fraction of the amount for graphic matter, and the labor force in the field could concentrate on sales and biographies rather than struggle to record the details of the cultural landscape on maps. At first, Andreas and Kett printed lists of all voters and taxpayers' names, augmented by "potted biographies" of subscribers inserted at the proper alphabetical place in the lists. In later volumes the non-subscribers were dropped.

The firm's first history was issued for De Kalb County, Illinois in late 1876, and over the following fifteen months ten others were published, all for contiguous counties crisscrossing northern Illinois. The use of voter lists as a base for the directory format of the biography section caused immediate problems. Despite instructions to canvassers to correct and update the lists, errors were frequent, as the *Rockton Herald* of Winnebago County, Illinois noted in December 1877:

It is surprising how the publishers could risk their reputation on a work which is manifestly a bundle of misstatements and blunders from end to end....Persons who have lived in this county twenty-five or thirty years have been wholly omitted....Names of persons have been slaughtered without mercy....Occupations have suffered in the hotchpotch manner of doing things. A farmer is converted into a builder...a doctor aspires to the sacred order of the priesthood. People have died before they were born, children have grown old while their parents were in tippets....

Fair or otherwise, such dissatisfactions, together with the firm's inability to control the quality and honesty of its agents, soured Kett on the business, and he sold his interest in the enterprise early in 1878. The firm was renamed the Western Historical Company, with A. T. Andreas as proprietor.

Once fully back in the saddle, Andreas tightened up his organization. He screened out and dismissed his less reliable canvassers, and revised the business so that historical material was prepared in advance and submitted to local figures for checking and possible correction. Thus reformed, Andreas pushed the publishing business hard by adhering to a strict formula. General sections on such topics as the U.S. Constitution, a brief history of the state and its laws, and

Maps for the Masses

"King of the Prairie" and "Lady Bushnell" boldly advertised the products of S.R. Hinkle's stock farm in Andreas's Fulton County, Illinois atlas of 1871.

various digests of census statistics were made up once and then reproduced whole in each separate county history. This left only historical sections on the county concerned, its townships and cities, and the biographical section to create anew. While the latter were without doubt substantial undertakings, the reuse of standard material sharply reduced the total effort for a given volume compared with the county atlases of old.

The Western Historical Company produced twenty-one histories of counties in Iowa, thirteen in Wisconsin, and one in Illinois between 1878 and 1881. In Iowa even more counties were covered under the imprints of the Iowa Historical Company, the Iowa Historical and Biographical Company, and The State Historical Company, all anchored in Des Moines, which from the physical similarity of their volumes appear to have been local fronts for Andreas's publishing activities in that state. In regions where counties were small or not populous, grouping of counties in single volumes was used and reached extensive proportions with Andreas's histories of Northern Wisconsin (1881) and Michigan's Upper Peninsula (1882). Clearly, it would be but a short step to engage in the compilation of a volume for an entire state—this time a state history rather than a state atlas. Recalling the publisher's mixed financial results with the latter, reaching for this again might have seemed fraught with risk.

To judge from the credit reports of the period, Andreas was once again on firm financial ground, as he had been during the purely county phase of his earlier atlas publishing career. The lower total investments required by county histories, together with the duplicate front material, turned out consistent profits. Even when Andreas took up the challenge of producing complete histories of Nebraska (1882) and Kansas (1883), the formula for economy and success did not desert him, and on the latter volume he made a clear profit of $20,000. Public reception of these volumes may have been little different from previous experience, but on balance it was positive,

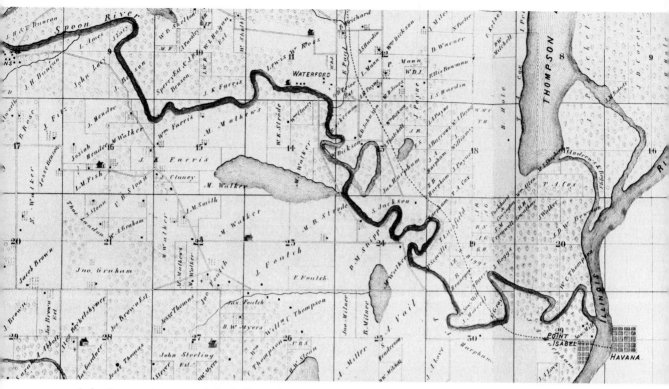

Andreas's county atlases commonly detailed houses, farmsteads, schools, waterways, and woodland cover. Here the Spoon River joins the Illinois at Point Isabel.

and the profits were there.

In addition to profit, civic responsibility probably entered into Andreas's decision to publish a multi-volume history of metropolitan Chicago. He had tested the waters of city history publishing with a successful volume on Milwaukee in 1882, and great care and planning went into the *History of Cook County, Illinois* (1884) as well as the projected three-volume city history. Andreas and his writers made considerable use of the archives of the Chicago Historical Society, to which he subsequently bequeathed a quantity of notes, correspondence, and preparatory material. No effort was spared in bringing out the *History of Chicago*, this monument to his adopted city, and he even went through bankruptcy after issuing volume one as a way of coping with the financial pressures. It took another two years to publish the remaining two volumes. The series has been recognized as a major achievement for the times.

The Twilight Years: 1886–1900

It is probable that the long, drawn-out publication of Andreas's *History of Chicago* so depleted his resources that he never was able to recover the old drive to organize grand new ventures.

Although there is some vague evidence that Andreas made money with the publication, he engaged in no significant publishing after 1886. In October 1882 he had joined the Illinois Commandery of the Military Order of the Loyal Legion of the United States. He had been buffeted by the vicissitudes of entrepreneurship, and as his energy flagged, stocktaking in the past and camaraderie among other men his age would have had its appeal.

This looking backwards also surfaced in his last known gainful employment. As he was bringing out the last of his Chicago volumes, Andreas sought his livelihood in the promotion and operation of panoramas devoted to dramatic Civil War battles. Panoramic exhibits were familiar fixtures in Chicago in the 1880s. Huge cylindrical life-sized paintings depicting great historical events in which the paying crowd could feel "in the thick of it" required large circular buildings, and the construction of these and their semi-permanent exhibitions called for organizational talents that Andreas certainly possessed. He was involved mostly in military panoramas: The Battle of Shiloh, the Monitor and Merrimac, The Battle of Bull Run, and The Battle of Gettysburg, but also—for good measure—the

Crucifixion. This activity doubtless kept him in the public eye through the late 1880s, for by 1891, as preparations were under way for the Columbian Exposition, he was appointed Secretary of the World's Fair Cooperative Bureau. At the time of the fair he also speculated heavily in hotels, dining places, and a transportation scheme and reportedly lost most of what fortune remained.

Following the fair little about his movements is known. After fifteen years living in comfortable single-family residences on Chicago's Near North Side, the *Chicago Blue Book* reported him boarding at the French House in Evanston in 1894–95. There is some evidence that he may have involved himself at this time, after a long hiatus, in the production of two landownership maps of Lake County and Stark County, Illinois, but accessible copies have not survived. Then he passed brief sojourns in a house in Highland Park (1896), boarding at the Windermere Hotel in Hyde Park (early 1897), followed by at least a year in Boston, boarding in a house on Beacon Hill. City directories in this period listed Andreas either as "publisher" or without occupation.

By 1899 Andreas had moved to New Rochelle, New York. What he did, if anything, as the century wound down is unclear. He died at the age of sixty-one on February 1, 1900 in New Rochelle. One obituary attributed his death to an old war wound in the right arm. He left behind Sophie, his wife of thirty-five years, who survived him another twenty, and two daughters, one of whom, Eulalia Lyter Andreas, went on to write one-act plays in early Hollywood. Although he was eligible, Andreas never drew a government pension. But when he left his wife with no source of income after his death, she was obliged immediately to seek a Civil War widow's pension.

Andreas's Map Publishing Achievement
The structure of the map trade in Chicago prior to Andreas's relocation in the city was relatively simple and fluid. The Great Fire served to accelerate developments already in motion and by creating a physical *tabula rasa* encouraged reconstruction of the printing and allied industries along more integrated lines. Andreas arrived in Chicago at the best possible moment to take part in the transformation and help shape it. While his direct and sustained involvement in map publishing was relatively short, he largely defined a specialized business approach to one segment of the trade that became institutionalized well into the twentieth century, and he is, therefore, a significant figure in the history of Chicago mapmaking.

Andreas came to Chicago with a specific goal—to publish illustrated county and state atlases; and, though not quite the first in this line, Andreas through his vision and drive eclipsed his colleagues and set the pace in atlas, and later county history publishing. This was a specialized enterprise; while producing atlases he engaged in no other types of mapmaking or marketing. Map publishers in other fields occasionally dabbled in county atlas work, but this was usually in later years and with different success. In the 1870s publishers of county atlases and histories were somewhat a breed apart, and the reason lies in the structure of the business. Most commercial mapmaking in a city like Chicago then was oriented towards trade sales through retail outlets with fixed locations, or upon contract sales to other businesses and institutions. County atlases, by contrast, were based upon subscription sales. Since these atlases, and the landownership maps that largely preceded them, were made for a contrived market, they had to be taken to the consumer, who was usually rural, dispersed, and distant. Therefore, the county atlas demanded an elaborate marketing system created each time for a given location and internal to the administrative structure of the atlas publishing firm. While this set the atlas apart from most other mapmaking, it gave it common ground with much general book publishing (dictionaries, bibles, encyclopedias) likewise dependent on sales by subscription.

Specialization opened the door to scale economies, and Andreas's regional atlas record (twenty-three county atlases) was a harbinger, along with the work of Frederick Beers in New York and David Lake in Philadelphia, of the true mass production of county atlases on a national scale that George A. Ogle would achieve in Chicago two decades later. The only other map publishing similarly organized was the urban real estate and later fire insurance atlas business, exemplified by Charles Rascher in Chicago, in which the system of frequent company-supplied map corrections bound publisher and consumer together

Maps for the Masses

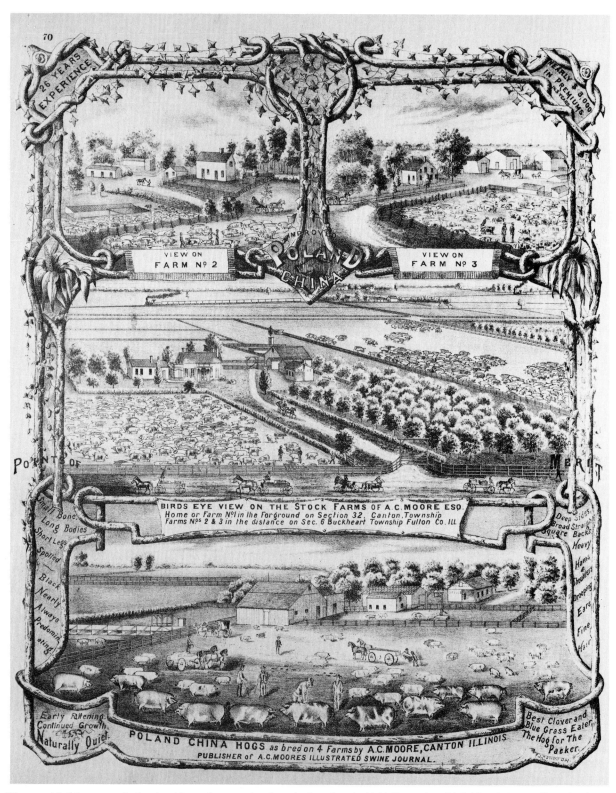

The potential of the county atlas for advertising one's own produce knew few bounds, as in this bird's-eye view of A.C. Moore's Canton, Illinois hog farm. "Small bone, long bodies, short legs, and drooping ears," evidently made Moore's improved Poland China breed "the hog for the packer."

in a long-term relationship. It was Andreas's strength that he understood the possibilities of the illustrated county atlas; it was his weakness that he overestimated them.

The Andreas achievement in the development of commercial cartography in nineteenth-century America was considerable. He seized upon the transition from county maps to atlases and created an atlas format designed to appeal to the broadest possible range of people, and in so doing democratized a genre of mapping traditionally confined to professional and cultivated classes and a scattering of administrative officials.

Andreas was not a craftsman with cartographic skills, nor was he a printer's artisan; he was a classic middleman, a salesman, a capitalist, and a preeminent manager. Like many other notable mapmakers he made things happen, but unlike Walling, the Thompsons, Warner, or Lake, he did not come to map publishing with a long apprenticeship in surveying and drafting, nor with the printer's experience of Robert Pearsall Smith, James McGuigan, or Edward Mendel. He surveyed the market and then assembled the technical expertise of others to produce the product he envisioned. While his capacity to organize the effort of others was reminiscent of Smith's own county map activity in what must, by 1876, already have seemed a simpler age, Andreas probably bore the greatest resemblance to his own one-time boss, friend, and competitor, Louis H. Everts. Also a salesman rather than a cartographer by innate talent, Everts was less visionary but more patient in fine-tuning his enterprise, and so lived to become a pillar of the Philadelphia publishing establishment.

These two men also represented something important occurring within the world of commercial cartography after the Civil War: the transfer of responsibility for both the content and the style, that is, the "look," of a genre of maps, passed from the cartographer and the printer to a third party, the capitalist-marketeer. This was hardly a general trend in the varied business of mapmaking, but it played a significant part in the broader shifts in the relations among mapmakers, as changes in techniques, economic and social demand for maps, and business organization redefined their traditional roles.

That Chicago emerged after the Civil War as a major center of American commercial mapmaking and as the premier center of subscription publishing in general was due in no small measure to the presence and drive of Andreas himself. His example was followed in Chicago by J. H. Beers and Company, Higgins, Belden and Company, O. L. Baskin and Company, the Kingman Brothers, and H. R. Page and Company, among others. George F. Cram turned out the odd county atlas and, through him, the prolific George Alden Ogle got his start in county atlas production in the late 1880s, not long after Andreas had ceased major publishing. But it was in his early prime that, with a combination of imagination, daring, and organizational panache, Alfred Theodore Andreas had found a curiously seductive way to bring maps to the common people.

For Further Reading

THE BEST INTRODUCTION to county maps and county atlases in the general context of American commercial mapmaking is Walter W. Ristow, *Maps for an Emerging Nation: Commercial Cartography in Nineteenth Century America* (Washington, D.C.: Library of Congress, 1977). A well-rounded sketch of the county atlas genre and its geographical spread over time is contained in Norman J. W. Thrower, "The County Atlas of the United States," *Surveying and Mapping*, 21 (1961): 365-73.

A highly entertaining, but not always objective, contemporary tirade against the 'vanity merchants' who published county atlases, including a specific discussion of Andreas's activities, is Bates Harrington, *How 'Tis Done: A Thorough Ventilation of the Numerous Schemes Conducted by Wandering Canvassers Together with the Various Advertising Dodges for the Swindling of the Public* (Chicago: Fidelity Publishing Co., 1879). A useful reconstruction of the use of newspapers in county atlas publishing by a man who once worked for Andreas is provided in Raymond and Betty Spahn, "Wesley Raymond Brink, History Huckster," *Journal of the Illinois State Historical Society*, 58 (Summer 1965): 117-38.

Primary materials essential to this study include an obituary, "Alfred Theodore Andreas," *Memorials of Deceased Companions of the Commandery of the State of Illinois, Military Order of the Loyal Legion of the United States* (Chicago, 1904): 493-96; and several manuscript credit reports in the R. G. Dun and Co. credit records, Manuscript Division, Baker Library, Harvard Business School, Boston. The relevant reports are contained in the following volumes: Illinois, Vol. 24, p. 172; Vol. 30, p. 103; Vol. 33, pp. 1b and 299; Vol. 36, p. 143; and Vol. 38, p. 171; and Scott County (Iowa) Vol. 1, pp. 224c, 329, 417, and 430.

The writer wishes to thank Walter W. Ristow, Arthur H. Robinson, and David Woodward for helpful comments on an earlier version of this essay.

Rand, McNally and Company in the Nineteenth Century: Reaching for a National Market

By Cynthia H. Peters

From modest beginnings as a railroad ticket printer, Rand, McNally and Company grew to a modern publisher of maps, atlases, and guidebooks. New production and marketing techniques were key to their success.

RAND, McNALLY AND COMPANY has become synonymous with mapmaking in American life, and yet comparatively little is known about the firm's rise to prominence in the nineteenth century. Rand, McNally's early success, from its quiet beginnings in 1856 until the early years of the twentieth century, is attributable both to good fortune and close calculation. In tracing this pattern of success, it is important to consider the production and marketing techniques employed in their map operations and the cost of Rand, McNally products in relation to the competition. Such a perspective clearly reveals how Rand, McNally managed to produce inexpensive maps that captured the mass market.

Rand, McNally's early development played an important role in the history of American publishing and printing and also in the history of Chicago. Their success highlights the movement of the American map publishing industry's center of gravity from the East to the Midwest. In contrast to the aesthetic quality of nineteenth-century European cartography, American traditions fostered a more utilitarian approach. The distribution and popularity of Rand, McNally maps were predicated upon this distinction.

The great advances made in commercial cartography in the United States during the second half of the nineteenth century were to a large extent an outgrowth of the railroad era, which demanded a new kind of map and mapping enterprise. As the new railroad network created a vast demand for maps, Chicago, its hub, became the logical location for a centralized map publication and distribution center. From modest beginnings as job printers of railroad tickets in the 1860s to their domination of railroad mapping by century's end, Rand, McNally shrewdly specialized in products required by this transportation revolution.

Ticket Printers Turn to Maps
William H. Rand and Andrew McNally acquired their early training in the printing business in Boston and Ireland respectively. Rand came to Chicago circuitously via Los Angeles, where he had established the city's first newspaper, *The Los Angeles Star*. In Chicago in June 1856, he opened a printing office over Keen and Lee's Bookstore at 148 Lake Street. McNally arrived in Chicago two years later and was hired by Rand. The first fifteen years of the venture were crucial. It was then that Rand and McNally decided to pursue job printing at the printing office of the *Chicago Tribune*. In May 1868, having formed the partnership of Rand, McNally and Company, they purchased the newspaper's interest in the job printing department. Railroad tickets and annual reports and stationery products soon became their most lucrative business.

The 1870s saw expansion and experimentation for Rand, McNally and securely established the company in the map field. Publication in July 1871 of the first issue of the *Western Railway Guide — The Traveler's Hand Book to All Western Railway and Steamboat Lines* served as a springboard eighteen months later for Rand, McNally's first maps. In the November 1872 *Railway Guide*, the company advertised: "Map Engraving a Specialty — All kinds of RELIEF LINE ENGRAVINGS promptly executed." The promised maps

---

Cynthia H. Peters is reference services librarian at the Newberry Library.

Rand, McNally and Company grew up at the height of the great railway age and shrewdly specialized in the products required by that transportation revolution: tickets, timetables, route maps, and (illustrated here) the comprehensive and ever-current Official Railway Guide.

National Map Market

appeared in the December issue along with the announcement that Rand, McNally was commencing high-quality mapmaking in earnest.

They used the new process of cerography or wax engraving. Although invented by Sidney Edwards Morse in the 1830s, cerography's full potential for cartography was not realized until the second half of the nineteenth century. Cerography combined two processes, relief engraving and electrotyping. After the base plate was coated with wax, the design was engraved into the wax, and the area between the incised lines was built up with additional wax. The plate then was electrotyped, producing a thin copperplate. The plate was reinforced with a backing of molten metal suitable for use on letterpress machines. The use of this process had several distinct advantages. Because the wax-engraved plate was more durable than copperplate, a single map could be produced in enormous editions for considerably lower costs. It also was easier to correct with revised geographical information. And finally, through the use of electrotyped plates, wax engraving had the added advantage of printing both type and graphics on the same sheet of paper.

Although others employed cerography before Rand, McNally entered the arena in 1872, none so fully exploited its potential. Rand, McNally's use of cerography was immediately profitable. Their first machine-colored railway map of the United States and Canada was published in 1874, the first pocket maps of Chicago and the Black Hills of South Dakota were published in 1875, the first edition of the *Business Atlas* was published in 1877, and their first local guidebook to Chicago was published in 1880. In addition to strictly cartographic works, Rand, McNally also began printing trade books as early as 1877 and school geographies in the 1880s.

The types of maps published by Rand, McNally between the time of their incorporation in 1873 and World War I illustrate their utilitarian and commercial approach to the mapping business. They always carefully designed their maps with the customer's needs in mind. Their approach to distribution, marketing, and pricing also was handled in a very pragmatic manner, as a look at

Railroad tickets and route maps constituted a large part of Rand, McNally and Company's early business. From the smallest short lines like the Sycamore and Cortland (left) to the great roads like the Chicago & North Western (right), the company captured the market.

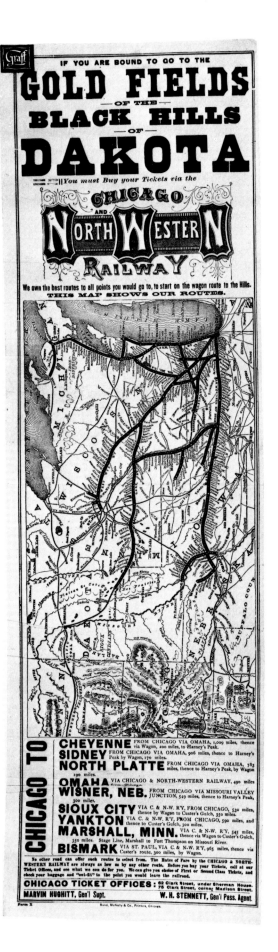

some of their products in relation to their distribution system shows.

One of the most significant publishing ventures that the company undertook during its formative years was the publication in 1876 of the *Business Atlas of the Mississippi Valley and Pacific Slopes*, which in 1877 became the *Business Atlas* and is now titled the *Commercial Atlas and Marketing Guide*. Issued annually ever since, it consists of large-scale maps of each state and territory printed in color and meticulously indexed. While Rand, McNally generally did not pursue original mapping for their *Business Atlas* or other cartographic works, relying instead upon government surveys and maps, they set a high standard for statistical accuracy. In order to obtain current census figures for towns and cities throughout the country that appeared on their maps, they routinely sent questionnaires and reply postcards to towns requesting the exact location and population of the community.

It was no less important that these early maps were printed on poor quality paper and had little aesthetic appeal: their purpose was strictly utilitarian. But the maps looked modern with clean and simple lettering, standardized coloring, and no superfluous decoration. In fact, the state maps in today's *Commercial Atlas and Marketing Guide* look very similar to the indexed pocket version of the 1890s. Serviceable maps in large quantities were the point—and the profit.

From its early history to the present, the company has remained a prolific publisher of affordable business, educational, and general family atlases. Between 1880 and 1917 Rand, McNally published some thirty different atlases, including the *Family Atlas of the United States*, the *Indexed Atlas of the World*, the *Library Atlas*, the *Pocket Atlas*, and the *Universal Atlas*. Most ranged in price from $1 to $12. There was more to their success, however, than convenient format and effective pricing. Shrewd advertising and distribution played an important role.

Rand, McNally was one of many publishing houses in Chicago that recognized the enormous potential of subscription sales, enabling buyers

Arriving in Chicago from Ireland in 1858, Andrew McNally (above) went into business with William H. Rand, himself a new arrival over a circuitous route from the East.

to purchase an atlas over an extended period of time. They handled subscription sales in various ways. The first method involved the direct sale of a specified number of atlases either whole or in parts as an incentive to subscribers. A slight variation involved farming out various atlases to newspapers or to Rand, McNally's subsidiary, the Continental Publishing Company in Chicago. For instance, in 1899, Rand, McNally increased the sales of their *Unrivaled Atlas* by publishing this work for various city newspapers such as the *Chicago Chronicle* and the *Philadelphia Public Ledger*, and placing their name prominently on it. The newspapers then sold the atlases to their subscribers.

Rand, McNally also realized early the potentially large market for pocket maps. In 1876, when the company sectionalized their *New Railroad and County Map of the United States and Canada* to form the basis for the *Business Atlas*, they began publication of a series of indexed pocket maps of individual states and territories ranging in price from 50 cents to $1. By the 1890s the cost for this series had been reduced to 25 cents per map, and another series, called vest pocket maps, was introduced at an even lower price of 15 cents. Throughout this period, Rand, McNally produced millions of pocket maps, in various sizes and formats, of every conceivable locale in the country.

Cerography made profitable such mass production of atlases; various forms of advertising helped to sell them. For example, the firm would send complimentary maps to newspaper offices all over the United States. In return, the individual newspapers usually would print Rand, McNally's enclosed blurb. The smaller midwestern towns favored this type of promotional advertising. *The Democrat* of Waukon, Iowa, in 1886 carried this notice:

We have received from Messrs. Rand, McNally & Co., of Chicago, a copy of their latest edition of an indexed county and township map and shippers guide of Iowa, a little work complete and authentic in all its features, showing in detail the complete railroad and express system of the state, accurately locating cities, towns, post offices, railway stations, villages, counties, etc. Every man having shipping transactions with the railway or express companies should send for a copy of this map. It costs but 25 cents and can be had on application to the publishers, 125-127 Monroe Street, Chicago.

Advertising in this way required a great deal of organization on Rand, McNally's part, but little investment, and it helped the company reach thousands of potential customers.

Private citizens were not the only buyers of Rand, McNally maps. While many of the firm's cartographic products were retailed in bookstores, in railway depots and aboard trains, and in newspaper offices, large quantities of maps also were purchased by railroad companies, which used them to encourage traffic and settlement along their routes. Such contracts with railroad companies greatly increased Rand, McNally's output of maps. Although the evidence of Rand, McNally's atlas and map press runs has not survived, it is fair to conclude that profits from their cartographic works published in large editions were considerable.

After successfully publishing all kinds of atlases and maps, guidebooks were the next logical extension of Rand, McNally's policy to specialize in one field of publishing and then diversify within that specialty. During the latter part of the nineteenth century, two important series of guidebooks appeared, "Sights and Scenes" and "Handy Guides," which offered guides for cities,

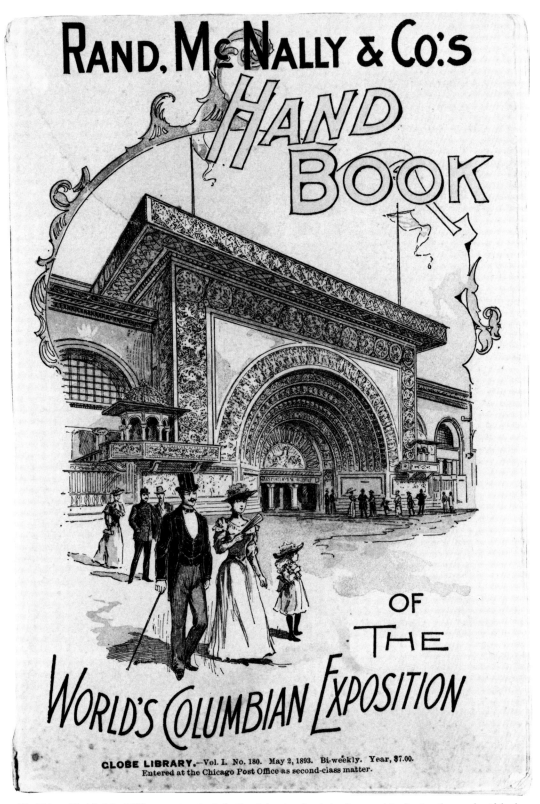

The Chicago World's Fair of 1893 was an opportunity for the city's mapmakers to produce a variety of maps, atlases, and guidebooks. Rand, McNally and Company's Handbook of the World's Columbian Exposition *is one example.*

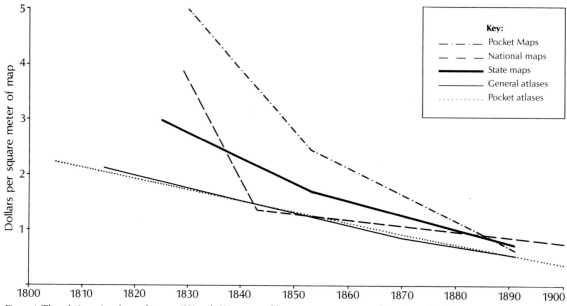

Figure 1. The relative price of maps between 1800 and 1900 compared by cost per square meter of map (a standard measure) reveals a dramatic drop from the earlier copperplate maps to the later wax engraved maps.

expositions, and vacation sites. They were illustrated with wood engravings, photographs, and, of course, maps. They were designed for a mobile population, and the railroads were again a prime avenue for creative marketing. Through retailing, advertising, subscription sales, and contracts with various private and governmental organizations, Rand, McNally created a powerful distribution apparatus that successfully captured a large segment of the market. By the end of the nineteenth century, the company was organized in six major divisions: tickets, railroad printing, map publications, bank publications, trade books, and educational textbooks. After World War I, highway maps gained a new and important place in the firm.

Map Prices: The Cerographic Edge

Rand, McNally's complex distribution system developed and matured because the company was able to produce inexpensive cartographic works. This was not always the case in mapmaking. Until the second half of the nineteenth century, map and atlas production costs remained high, ruling out any prospect of mass production or large-scale marketing. Gradually, as new printing technologies reduced the cost of production, mapping companies were able simultaneously to increase output and reduce prices for the consumer. This was the general trend after 1820. A brief comparative examination of the costs of commercial map reproductions during this period illustrates the changes occurring in the industry and how Rand, McNally profited from them.

Before the 1830s, copperplate engraved maps, from such American mapping establishments as Henry S. Tanner's and John Melish's, dominated the market and remained expensive. By the 1840s and 1850s many mapping companies, such as those of Joseph Hutchins Colton and Samuel Augustus Mitchell, used copper and steel plate engravings in combination with lithography. This method drastically reduced the cost of their maps. By the 1870s, however, high-quality maps had been superseded by maps printed from wax engravings, a process even less expensive than lithography. Rand, McNally and George F. Cram dominated this field.

A close look at retail prices for maps and trends in personal income measured by average wages in the nineteenth century offers some insight into how the map market changed from 1800 to 1900. A survey of representative types of maps—sheet maps, pocket maps, atlases, and pocket atlases—reveals a marked general decline in map prices and, more important, a dramatic drop in map costs per square meter (a standardized measure over time) from copperplate engraved maps to wax-engraved maps (figure 1). Though most of the examined copperplate engraved maps ranged from $2 to $4 and lithographed maps

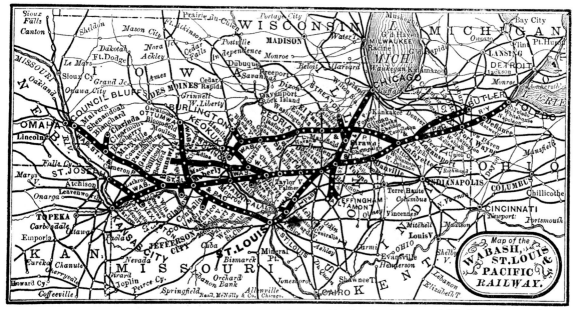

Appearing in advertising, company timetables, and in the Official Railway Guide, *the railroad route map boldly depicted a new and sometimes distorted vision of geographical space and the means meant to abridge it.*

from $1 to $2 in cost per square meter, all of the Rand, McNally maps studied were considerably under $1. During the early part of the century the cost per square meter of map varied widely with the type of map produced. The cost, then, was a function both of the printing process and other factors, such as research for compilation, size of press runs, and marketing techniques employed—all of which affected the final retail price of the map. But by the latter part of the century, Rand, McNally had achieved a nearly uniform cost per square meter for all of their cartographic publications. Thus it appears that the principal cost factor was quite directly the printing process itself, cerography, which was extremely inexpensive.

The other critical factor in calculating a product's affordability is the cost of living. Between 1800 and 1900 the trend in the reconstructed consumer price index was a gradual decline which occurred in two cycles. And while the cost of living fell in the nineteenth century, the wage levels in the United States rose significantly. For instance, a non-farm laborer earned 75 cents a day in 1830, and by 1899 that same laborer earned $1.41 a day. In that seventy-year period, the wages of most workers in the United States nearly doubled while their cost of living dropped by nearly one-quarter. Rand, McNally maps, therefore, were relatively cheaper in 1900 than were Mathew Carey's maps in 1800, and clearly more affordable. Indeed, the cost of maps declined at a greater rate than the decline of consumer prices generally.

Financial Success of Rand, McNally

Considering that Rand, McNally and Company was participating in a market which emphasized mass production, innovative advertising, wide distribution, and competitive pricing, this company did well to be financially prosperous from the beginning. Success is evident in their growth figures and assets. After incorporation on June 1, 1873, sales and profits dramatically increased over the next forty years. According to typescript copies of the "Board Minutes of Stockholders' Meetings" and the "Treasurer's Reports," Rand, McNally was an immediate success. The firm's sales show a continual increase from approximately $426,000 in 1877 to $2,000,000 in 1913. That is, the company's sales during this thirty-six-year period had increased by 375 percent, a very impressive figure.

Historically, control remained tightly in the hands of a limited number of stockholders. In 1873 when the company incorporated, 2,000 shares were distributed principally among the four board members: William H. Rand, Andrew McNally, Henry Kidder, and George Poole. When Rand retired in 1899, he sold his shares in the company to the remaining stockholders. By 1904 Andrew McNally had become the principal stockholder with 5,715 shares; followed by John R. Walsh, 1,000 shares; Fred M. Blount, 110 shares; and Fred McNally, 106 shares. Rand, McNally's management of their assets was one of the keys to their success. Equally as important was their ability to cope with the Chicago Typographical Union. Throughout the late nineteenth and early twentieth centuries, the printing unions continued to press for higher wages, a shorter work week, and better benefits. In reaction to their demands, the leading proprietors of publishing firms, including Rand, McNally, came together to organize the Chicago Typothetae in 1887 and the Master Printers' Association in 1899. Although Rand, McNally opposed the formation of stronger unions in the printing industry, they remained one of the better employers in Chicago. According to the Manufacturers Census for Illinois in 1880, Rand, McNally generally paid high wages ($3 per day for a skilled mechanic and $1.20 per day for an ordinary laborer). Eventually though, as union pressure increased in the early twentieth century, the firm sought to control labor costs by establishing printing plants in rural areas where open-shop policies often existed.

The company's early and continued success is evident also in the rapid expansion of their facilities. Only seven years after the company's incorporation, they erected a six-story building at 125–127 West Monroe Street, which they occupied from 1880 to 1889. Their tenants included the Chicago, Milwaukee & St. Paul Railroad and the World's Fair management. By the end of the 1880s, they had outgrown this building and constructed a new ten-story, all-steel building at 166–168 Adams Street. Designed by Burnham and Root, the new building cost $450,000. They moved again to new quarters at 536 South Clark Street in 1912, and once more in 1952 to Skokie. Moving into the wider world of business, Rand, McNally not only reinvested profits in their own company, but diversified into related industries such as publishing houses and railroads. As a means of improving their distribution network, for example, in 1884 Rand, McNally formed the subsidiary, Continental Publishing Company, with a capital stock of $200,000 and Andrew McNally as president. In addition, they invested in the Atchison, Topeka & Santa Fe Railroad, the Southern Indiana Railway, and the Chicago and Alton Railroad.

The history of Rand, McNally and Company exemplifies not only common characteristics of American commercial enterprise, but also reveals the impact that important social trends had on map publishing. The diffusion of geographical knowledge was probably the company's most important contribution to American life in the nineteenth century. Their use of cerography permitted huge production runs, and consequently geographical knowledge was dispensed cheaply and widely. Their organizational framework—printing, publishing, and mapmaking—allowed various specializations to be pursued from beginning to end. As the forerunners of today's information specialists, Rand, McNally in the nineteenth century met the demands for geographical knowledge in an era of rapid change. Their success ensured a longevity that is remarkable in modern mapmaking.

For Further Reading

A FULL-LENGTH scholarly study of the history of Rand, McNally and Company has yet to be written. This essay draws from primary research in the Company's archives in Skokie, Illinois. In addition to large map, atlas, and manuscript holdings, the archive contains an unpublished document by Bruce Grant, "Notes on the History of Rand, McNally and Company," written in the 1950s in anticipation of the firm's centennial anniversary, all of which proved most helpful.

Maps produced by the company and its competitors also were examined in the collections of the Library of Congress, the Newberry Library, and Regenstein Library at the University of Chicago. Further primary evidence was gleaned from the Illinois manuscript schedules of the 1880 U.S. Census of Manufactures held in the State Historical Library in Springfield, and various mapmakers' sales catalogs such as *Rand, McNally's Descriptive Catalog of Maps, Atlases, Globes, etc.* (Chicago, 1891).

Useful secondary sources include Andrew McNally III, "Rand, McNally in the World of American Cartography," *American Cartographer* 4 (1977): 101-10; *Ranally World*, 7 numbers (Chicago: Rand, McNally and Co., 1956); Emily Clark Brown, *Book and Job Printing in Chicago* (Chicago: The University of Chicago Press, 1931); Andrew M. Modelski, comp., *Railroad Maps of the United States* (Washington, D.C.: Library of Congress, 1975), David Woodward, *The All-American Map* (Chicago: The University of Chicago Press, 1977); and Stanley Liebergoff, "Wage Trends, 1800-1900," in *Trends in the American Economy in the Nineteenth Century* (Princeton: Princeton University Press, 1960).

Made In Chicago: Maps And Atlases Printed In Chicago Before The Fire

Compiled by Robert W. Karrow, Jr.

This list includes separately published maps, maps in books, and atlases printed in Chicago before 1872. In the interest of space, views have been excluded, as have maps which bear a Chicago imprint (e.g. "Published by Rufus Blanchard, Chicago") but which are known, or strongly suspected, to have been printed elsewhere. The list was derived from the *Checklist of Printed Maps of the Middle West to 1900* (Boston: G. K. Hall & Co., 1981), a cooperative catalog of some 26,600 titles in midwestern institutions and the Library of Congress.

The *Checklist* includes many maps of North America and the United States, in addition to those of the midwest and its individual states, counties, cities, and towns. While maps of many areas of the country were thus systematically excluded (a map of California printed in Chicago, for example, would not appear in the *Checklist* or, consequently, in this list), the evidence suggests that the number of maps produced for an area drops off rapidly as the distance from Chicago increases, and the pre-fire maps of areas outside the bounds set by the *Checklist* cannot have been numerous.

Each entry in this list includes a brief title followed by the name of the compiler (if any) set off by a slash. Additional names of persons or firms, set off by periods, are of Chicagoans associated with the map as publishers, printers (usually lithographic printers), engravers, colorists, or a combination of these functions. The number following an entry refers to the *Checklist*, where a fuller description of the item will be found. A few maps identified since the publication of the *Checklist* and not included in it have been added to this list; these entries have a library abbreviation in lieu of a number. The library abbreviations are: BL = British Library, CHS = Chicago Historical Society, IHS = Indiana Historical Society, NL = Newberry Library, and PRI = Private collection. Other abbreviations used are "lith." for lithographer, "eng." for engraver, and "sc." for the Latin *sculptor* or engraver. An asterisk (*) preceding a map title indicates that the dating is conjectural.

It is a pleasure to acknowledge the assistance of Michael P. Conzen, Grant Dean, and Patricia A. Moore in the preparation of this list.

1844
Chicago in 1812. In Juliette Kinzie, *Narrative of the massacre at Chicago.* 4 0286

1845
Chicago in 1812. In *Business advertiser and general directory of the city of Chicago.* CHS

1847
[*Map of the route of the Galena and Chicago Union Railroad*]. R. N. White, sc. In Richard P. Morgan, *Report of the survey of the route of the Galena and Chicago Union Rail Road.* 4 1500

1852
Map of the city of Chicago / A. H. & C. Burley. In *Chicago city directory.* 4 0320-21

1853
CHICAGO AND VICINTY
The city of Chicago. H. Acheson lith. 4 0323

Map of Chicago and its southern & western suburbs. Ed. Mendel lith. 4 0326

Map of the city of Chicago / A. H. & C. Burley. In *Chicago city directory.* 4 0324

ILLINOIS
Map of the Racine, Janesville & Mississippi Rail-Road and its connections. Acheson & Rodway lith. 6 1724

Map showing the Galena & Chicago Union Rail Road and its connections. Mendel & Atwood lith. In its *Sixth annual report.* 4 1530

1854
CHICAGO AND VICINITY
The city of Chicago. H. Acheson lith. 4 0330

J. B. F Russell & Co.'s map of the city of Chicago. H. Acheson lith. 4 0331

Map of the city of Chicago / A. H. & C. Burley. From a city directory. 4 0335

N. P. Iglehart & Cos. map of the city of Chicago. H. Acheson lith. 4 0332

Rees & Kerfoot's Chicago rail-road map. H. Hart. H. Acheson lith. In *Rail-roads, history and commerce of Chicago.* 1 1039

Rees & Kerfoot's map of the city of Chicago. H. Acheson lith. 4 0333

ILLINOIS
D. B. Cooke & Co.'s railway guide for Illinois. Ed. Mendel lith. 4 1540

Galena and Chicago Union Rail Road. H. Acheson lith. In its *Seventh annual report.* 4 1541

Map of Illinois. Rufus Blanchard. Ed. Mendel lith. In *Railroads, history, and commerce of Chicago.* 4 1533

Morse's map of Illinois. Rufus Blanchard. 4 1537

OTHER
Map showing the position of the Lyons, Iowa, Central Railroad. H. Acheson lith. In its *First annual report.* 8 0693

1855
CHICAGO AND VICINITY
Map of Illinois Central Railroad Company's depot grounds & buildings in Chicago. Ed. Mendel lith. 4 0347

New map of Chicago. Ed. Mendel lith. 4 0343-46

Plan no. 1, Chicago sewerage. Ed. Mendel lith. In *Board of Sewerage Commissioners Report.* 4 0348

ILLINOIS
City of Springfield, Sangamon Co., Ills. Chicago Printing Co. Ed. Mendel lith. In *Springfield city directory, and Sangamon County advertiser for 1855-6.* 4 2693

D. B. Cooke & Co.'s railway guide for Illinois. H. Acheson lith. 4 1552

Map of Elgin, Kane County, Illinois / Thomas Doran. Ed. Mendel lith. CHS

Map of the state of Illinois N. America / Henry Greenbaum & T. W. Sampson. Ed. Mendel lith. 4 1547

Railway guide for Illinois. 4 1554

Sectional maps, showing 2,500,000 acres, farm and wood lands, of the Illinois Central Rail Road Company. H. Acheson lith. 4 1553

OTHER
D. B. Cooke & Co.'s rail way guide for Indiana. H. Acheson lith. 3 0547

D. B. Cooke & Co.'s railway guide for Ohio. CHS

**Map of Dunnville, Dunn County, Wisconsin* / Miller & Denniston. Ed. Mendel lith. 6 0338

Map of Kansas & Nebraska. Ed. Mendel lith. In Walter B. Sloan, *History and map of Kansas & Nebraska.* 13 0543

**Map of Kansas and Nebraska Territories.* Mellen & Co. Ed. Mendel lith. 13 0546

Map of Mineral Point [Wis.] / Charles Temple. J. J. O'Shannessy Lith. 6 1002

**Map shewing the connection of Chicago and other lake ports with northern and central Wisconsin by rail road.* H. Acheson lith. 6 1754

Morse's map of Indiana. Rufus Blanchard. 3 0546

Sectional map of southern Minnesota / S. Holsteen. Ed. Mendel lith. 7 1256

1856
CHICAGO AND VICINITY
Chicago harbor & bar / J. D. Graham. Ed. Mendel lith. 4 0786-91

This 1856 map and others (same title) dated 1857 and 1858 accompanied the reports of Graham, an Army engineer.

Gager's city directory map of Chicago. Ed. Mendel lith. 4 0349

Made in Chicago

ILLINOIS
Map of Morris, Grundy County, Illinois / Thomas Doran. Ed. Mendel lith. PRI
Map of Sheffield and Savannah rail-road / W. G. Wheaton. H. Acheson lith. In Wheaton, *Report upon the preliminary survey.* 4 1562
**Map showing the relative position of Rock Island with the north western states.* Ed. Mendel lith. 4 2553

OTHER
Black Lake Harbor, Michigan / John R. Mayer. Ed. Mendel lith. 5 1261
Grand River Harbor...Michigan / John R. Mayer. Ed. Mendel lith. 5 1054
Map of the Garden City mines & vicinity [*Mich.*]. Ed. Mendel lith. In Garden City Mining Co., *Articles of incorporation.* 5 0990
Medford, Steele County, Minnesota / Ozro A. Thomas. Ed. Mendel lith. 7 0549
St. Joseph Harbor, Michigan / John R. Mayer. Ed. Mendel lith. 5 3497

1857
CHICAGO AND VICINITY
Chicago, the largest primary grain port in the world. H. Acheson lith. 4 0352
Map of Chicago. Rufus Blanchard. 4 0353-54
Plan showing sewers laid...to end of 1857. From a book. 4 0355
[*South Branch of the Chicago River in sec. 29*]. Ed. Mendel lith. Map on letterhead. 4 0978

ILLINOIS
D. B. Cooke & Cos. miniature rail road map of Illinois. J. Gemmell lith. 4 1574
**Design for the grounds of the State Normal School, Bloomington, Illinois.* Ed. Mendel lith. 4 1897
Map of Bureau County, Illinois / N. Matson. Ed. Mendel lith. 4 0153
Map of Tazewell County, Illinois / Thos. King, Jr. Reen & Shober lith. 4 2732
Map of the present & prospective rail-road connections of the city of Freeport. J. Gemmell lith. In Boss & Burrows, *Present advantages and future prospects of the city of Freeport.* 4 1234

OTHER
*Franklin City [*Neb.*]. J. Gemmell lith. 12 0181
Kenosha Harbor, Wisconsin / John R. Mayer. Ed. Mendel lith. 6 0638
Map of a portion of Douglas & La Pointe counties, Wisconsin / Mitchell, Rice & Relf. Ed. Mendel lith. 6 0326
**Map of Grand-Haven, Ferrysburg, and MillPoint, Ottawa County, Michigan* / Sam S. Montague. Ed. Mendel lith. 5 1044
Map of Grant County, Wisconsin / J. Wilson, Jr. Ed. Mendel lith. 6 0488
Map of Scott Co. Iowa and Rock Island Co. Illinois / C. H. Stoddard, Ed. Mendel lith. 8 1736
Mouth of South Black River, Michigan / John R. Mayer. Ed. Mendel lith. 5 0269
Map of the city of Davenport and its suburbs, Scott County, Iowa / James T. Hogane & H. Lambach. Ed. Mendel lith. 8 0327
Map of Winona [*Minn.*] / Lyman G. Bennett. J. J. O'Shannessy lith. 7 1950
Mouth of Kalamazoo River, Michigan / John R. Mayer. Ed. Mendel lith. 5 1510

1858
CHICAGO AND VICINITY
Chicago. John Gemmell lith. 4 0359
New map of Chicago comprising the whole city / N. P. Iglehart & Co. Ed. Mendel lith. 4 0360

ILLINOIS
Diagram of the state of Illinois / H. A. Ulffers. Ed. Mendel lith. In J. G. Norwood, *Abstract of a report on Illinois coals.* 4 1575
Larrance's post office chart [*of Illinois*]. Ed. Mendel lith. 4 1576
Map of a portion of Henry County, Illinois. Ed. Mendel lith. 4 1367
Map of the Illinois River Rail Road. Ed. Mendel lith. From its *Chief Engineer's report.* 4 1577
Map of the Jacksonville and Savannah R. R. / H. J. Vaughan. J. Gemmell lith. From its *Engineer's report.* 4 1578
*Map of Tuscola [*Ill.*]. Reen & Shober lith. 4 2758

OTHER
**Map of Blue Earth County, Minnesota* / C. C. Whitman. Ed. Mendel lith. 7 0048
Map of Fillmore County, Minn. / J. W. Bishop. Ed. Mendel lith. In Bishop, *History of Fillmore County.* 7 0307
Map of Iowa showing the line of the Iowa Central Air Line Rail Road. Ed. Mendel lith. In its *First annual report.* 8 0731
Map of Rock County, Wisconsin / A. B. Miller & Orrin Guernsey. J. Gemmell lith. 6 1253
Map of the property of New Diggings & Shullsburg Mining Company [*Wis.*] / Thos. Daniel. Ed. Mendel lith. 6 1058
**Plan of Houghton, La Pointe Co., Wisconsin* / G. L. Brunshweiler. Ed. Mendel lith. 6 0557
**Railroad map of Wisconsin.* Rufus Blanchard. 6 1794

1859
CHICAGO AND VICINITY
**Chicago...for D. B. Cooke & Co.'s directory.* 4 0363

ILLINOIS
Map of Hancock County, Illinois / Holmes & Arnold. Chas. Shober lith. 4 1339
Map of Joliet, Will Co., Illinois / A Veith. Ed. Mendel lith. 4 1924

OTHER
City of Ypsilanti, Washtenaw County, Michigan / C. S. Woodard. Chas. Shober lith. 5 3900
**Map of Chatfield, Minn.* / J. W. Bishop. Ed. Mendel lith. 7 0123
Map of Linn County, Iowa / McWilliams & Thompson. Ed. Mendel lith. 8 1361
**Map of Warren County, Iowa* / Dan A. Poorman. Ed. Mendel lith. 8 1873
W. B. Horner's railway & route map to the gold regions in Nebraska and Kansas. J. Gemmell lith. In Horner, *Gold regions of Kansas and Nebraska.* 12 0372
Wood's map of Milwaukee. Chas. Shober lith. 6 0904

1850s
CHICAGO AND VICINITY
Blanchard's map of Chicago. Rufus Blanchard. 4 0322
Stone & Whitney subdivision. Ed. Mendel lith. 4 0963

ILLINOIS
Map of the Illinois River Railroad / W. G. Wheaton. J. Gemmell lith. 4 1511

OTHER
Omaha Nebraska. Ed. Mendel lith. 12 0962
Sectional Nebraska and Kansas / James M. Lowe. Rufus Blanchard. J. Gemmell lith. 12 0357
The village of Beloit, Rock Co., Wisconsin / S. T. Merrill. H. Acheson lith. 6 0108

1860
CHICAGO AND VICINITY
Chicago...for D. B. Cooke & Co.'s directory. J. Gemmell eng. 4 0364

Kerfoot's map of Chicago. 4 0365

ILLINOIS
Map of De Kalb County, Illinois / Daniel W. Lamb. Chas. Shober lith. 4 1080
Map of Henry County, Illinois / P. Holmes. Ed. Mendel lith. 4 1368

OTHER
Map of Dodge County, Wisconsin / F. Hess. S. H. Burhans & C. G. Scott. Ed. Mendel lith. 6 0310
**Map of LaFayette Co.* [*Wis.*] / Warren Gray. Chas. Shober lith. 6 0708
**Map of Muskegon River* [*Mich.*] / Wm. P. Innes. Ed. Mendel lith. 5 2986
Map of the Cedar Rapids & Missouri River Rail Road. Chas. Shober lith. In its *First report.* 1 1214
Sketch of the quartz mining region of the Rocky Mountains / S. W. Burt & E. L. Berthoud. Ed. Mendel lith. 13 0594

1861
CHICAGO AND VICINITY
Map of Cook County, Illinois / W. L. Flower. S. H. Burhans & J. Van Vechten. Ed. Mendel lith. 4 1029

ILLINOIS
Map of Green County, Illinois / Charles R. Arnold. Ed. Mendel lith. 4 1306
Map of Knox County, Illinois / M. H. Thompson. Ed. Mendel lith. 4 1967
Map of Stark Co., Ill. / M. H. Thompson. Chas. Shober lith. BL
Thompson's district map of Illinois 1861, showing congressional, senatorial & legislative districts. Chas. Shober lith. 4 1589

OTHER
Geological diagram of the Garden City Mining Co.'s location [*Mich.*] / Henry Merryweather. Ed. Mendel lith. 5 0991
Map of Dane Co., Wis. / A. Ligowski. Ed. Mendel lith. 6 0267
Map of Ionia County, Michigan / Geo. W. Wilson. Ed. Mendel lith. 5 1372
Map of Noble Co., Ind. / E. B. Gerber. Ed. Mendel lith. 3 1093
**Map of the Chicago & North Western Railway & its connections.* Chas. Shober lith. 6 1804
**Map showing the position of Chicago, in connection with the North West.* Ed. Mendel lith. 1 0888

1862
CHICAGO AND VICINITY
Blanchard's guide map of Chicago. 4 0374
Map of Chicago / P. T. Sherlock. Ed. Mendel lith. 4 0376
Map of Cook County, Illinois / W. L. Flower. S. H. Burhans & J. Van Vechten. Ed. Mendel lith. 4 1030
Map of the business portion of Chicago / E. Whitefield. Rufus Blanchard. 4 2082
New counting-house map of Chicago / S. H. Kerfoot & Co. 4 0375

ILLINOIS
Map of McHenry Co., Ill. / M. H. Thompson & Bro. Chas. Shober lith. 4 2136
Map of Will County, Illinois / S. H. Burhans & J. Van Vechten. Ed. Mendel lith. 4 2886

OTHER
Map of Cass Co., Ind. / R. J. Skinner. Ed. Mendel lith. IHS
Map of Fond du Lac County, Wisconsin / W. T. Coneys, M. L. Bogert. Ed. Mendel lith. 6 0412
Map of the Chicago & Northwestern Railway & its connections. Chas. Shober lith. 1 0891
Map of Winnebago County, Wisconsin / G. A. Randall & C. Palmer. Ed. Mendel lith. 6 1613

1863
CHICAGO AND VICINTY
Chicago, drawn from Davie's atlas / W. L. Flower & J. Van Vechten. Chas. Shober lith. 4 0377

Map of Chicago / J. Van Vechten. Chas. Shober lith. 4 0380

Map of Chicago...for Halpin and Bailey's city directory. J. Van Vechten. Chas. Shober lith. 4 0378-79

ILLINOIS
Blanchard's township map of Illinois. Chas. Shober lith. 4 1596

Map of Champaign County, state of Illinois / Alex. Bowman. Chas. Shober lith. 4 0262

Map of Grundy County, Ill. / Thomas Doran. Ed. Mendel lith. PRI

Map of Lee County, Illinois / Joseph Crawford & Jason C. Ayres. Ed. Mendel lith. 4 2016

Map of Marshall & Putnam Cos., Ill. / M. H. Thompson. Chas. Shober lith. BL

A New township map of Illinois / C. D. Wilbur. Chas. Shober lith. 4 1601

OTHER
The coast pilot chart of Lake Michigan. Chas. Shober lith. In James Barnet, *Coast pilot for the lakes.* 1 0243

Guide map of the city of Detroit / Eugene Robinson. Ed. Mendel lith. 5 0603

Map of Carroll Co., Ind. / R. J. Skinner & Bennet. Chas. Shober lith. NL

Map of north eastern Iowa and south eastern Minnesota / W. J. Barney. Ed. Mendel lith. 8 1216

Map of St. Joseph County, Indiana / M. W. Stokes. Chas. Shober lith. 3 1266

1864
CHICAGO AND VICINITY
Blanchard's guide map of Chicago. 4 0382

Map of Chicago...for Bailey's city directory. Chas. Shober lith. 4 0383

ILLINOIS
Illinois 1864 / Walter S. Frazier. Chas. Shober lith. 4 1582

Map of Logan Co., Ill. / S. H. Burhans & L. M. Snyder. Ed. Mendel lith. BL

Miniature map of Illinois. Chas. Shober lith. CHS

Van Vechten's congressional map of Illinois. Chas. Shober lith. 4 1609

Van Vechten's representative map of Illinois. Chas. Shober lith. 4 1610

Van Vechten's senatorial map of Illinois. Chas. Shober lith. 4 1611

Wilbur's miniature map of Illinois. Chas. Shober lith. 4 1607

Wilbur's physical and descriptive map of Illinois. Chas. Shober lith. 4 1608

OTHER
Map of Monticello, Jones Co., Iowa / W. G. Hammond. Ed. Mendel lith. 8 1505

New & correct sectional map of the iron, silver, lead & gold region, Lake Superior / J. A. Banfield. Ed. Mendel lith. 5 1841

1865
CHICAGO AND VICINITY
Fire insurance maps of the city of Chicago / Frederick Cook. B. W. Phillips & Co. Ed. Mendel lith. 4 0387

Map of Chicago. Ed. Mendel lith. In *Chicago, a strangers' and tourists' guide.* 4 0385

Van Vechten's official guide map of Chicago. Chas. Shober lith. 4 0386

ILLINOIS
Map of Menard Co., Ill. / L. M. Snyder. Chas. Shober lith. BL

Sterling, Whiteside Co., Ills. / Nathan Hicks. H. Acheson lith. CHS

Wilbur's miniature map of Illinois. Chas. Shober lith. 4 1612

OTHER
Cabinet map of Minnesota. Rufus Blanchard. 7 0763

Map of Lincoln, Mason Co., Mich. Chas. Shober lith. 5 1683

New sectional & township map of the state of Wisconsin / G. A. Randall. Ed. Mendel lith. 6 1819

Sectional map of Kansas / Geo. O. Willmarth. Rufus Blanchard. 13 0610

1866
CHICAGO AND VICINITY
Great auction sale of choice grove property, around the Douglas monument. Clarke, Layton & Co. 4 1107

Guide map of Chicago. Rufus Blanchard. In his *Citizen's guide for the city of Chicago.* 4 0389

Plan of Lincoln Park, Chicago / Swain Nelson. L. Nelke lith. 4 2047

ILLINOIS
Blanchard's township map of Illinois. Chas. Shober lith. 4 1613

OTHER
Blanchard's map of Iowa and Nebraska. 8 0762

Blanchard's map of Wisconsin and northern Michigan. Chas. Shober lith. 6 1823

Blanchard's map of the north western states. Rufus Blanchard. 1 0898

Map of LaFayette Co [Wis.] / Warren Gray. Chas. Shober lith. 6 0709

Map of Vermilion Lake, St. Louis Co., Minnesota / J. Hill, Western Engraving Co. 7 0470

Plan of Hamilton, Scott County, Minnesota. Ed. Mendel lith. 7 1715

1867
CHICAGO AND VICINITY
Blanchard's 1867 map of Chicago and environs / Otto Peltzer. 4 0399

Blanchard's guide map of Chicago. In his *Citizen's guide for the city of Chicago.* 4 0392-93

Great auction sale of property near Union Park / Clarke, Layton & Co. Landon & Kroff printers. 4 2329

Guide map of Chicago, engraved for N. S. Davis, M. D. 4 0395

Map, city of Chicago, expressly engraved for W. S. Spencer. Chas. Shober lith. 4 0400

[Map of Chicago] Baird & Bradley. 4 0391

Map of Chicago. Ed. Mendel lith. In *Guide to the city of Chicago.* 4 0396

Rufus Blanchard's guide map of Chicago, engraved for Richard Edwards' new directory. 4 0394

ILLINOIS
Blanchard's township map of Illinois. Chas. Shober lith. 4 1618

Map of Bureau County, Illinois / N. Matson. Tribune Co. 4 0154

Map of Vermillion Co., Ill. / A. Bowman. Chas. Shober lith. BL

OTHER
Blanchard's map of Iowa and Nebraska. 8 0772

Blanchard's map of Minnesota and Dakota. 7 0772

Blanchard's map of Missouri and Kansas. Chas. Shober lith. 9 0906

Blanchard's map of the north western states. Rufus Blanchard. Shober & Co. lith. 1 0899-900

Map of Ramsey and Manomin Counties / L. G. Bennett. Chas. Shober lith. 7 1805

Map of Ramsey County, Minnesota / L. G. Bennett. Chas. Shober lith. 7 1447

Map of the northwest in 1867. Baker eng. In George Gale, *Upper Mississippi.* 1 0901

Map of Winona County, Minnesota / L. G. Bennett & A. C. Smith. Chas. Shober lith. [Actually an atlas]. 7 1965

1868
CHICAGO AND VICINITY
Blanchard's map of Chicago and environs / Otto Peltzer. Chas. Shober lith. 4 0404

Blanchard's map of Cook and DuPage Counties. 4 1031

Great auction sale of Egandale property / Clarke, Layton & Co. Horton & Leonard printers. 4 1402

Guide map of Chicago. Rufus Blanchard. From *Citizen's guide for the city of Chicago;* also in John S. Wright, *Chicago past, present, future.* 4 0401-02, 04

Guide map of Chicago / Geo. F. Cram. Chas. Shober lith. 4 0405

Hyde Park. Ed. Mendel lith. 4 0926

John Wentworth's subdivision. Horton & Leonard printers. 4 1108

Map of Chicago / J. Van Vechten. Chas. Shober lith. 4 0407

Map of land of Chicago Mutual Homestead Association. Baker eng. Horton & Leonard printers. 4 0408

Official guide map of Chicago. J. Van Vechten. Chas. Shober lith. 4 0406

Sectional map of Cook County, Ill. / John H. Hewitt. Chas. Shober lith. 4 1032

ILLINOIS
[Map of DeKalb County, Ill.] In Henry L. Boies, *History of DeKalb County.* 4 1081

Map of Marseilles at the Grand Rapids of the Illinois river / J. Gray & S. B. Carter. BL

Map of Mason Co., Ill. / L. M. Snyder. Chas. Shober lith. BL

Sectional map, Knox County, Illinois. Chas. Shober lith. In Dewey's *County directory.* 4 1968

OTHER
Blanchard's map of Wisconsin and northern Michigan / Blanchard & Cram. Chas. Shober lith. 6 1831

Cabinet map of the United States and territories. Rufus Blanchard. 1 1678

Cabinet map of the western states. Rufus Blanchard. 1 0903

Map of Iowa. Rufus Blanchard. Chas. Shober lith. 8 0779

Map of Missouri and Kansas. Rufus Blanchard. 9 0909

Map of Porter Co., Ind. / Simon Murphy. Chas. Shober lith. BL

New railroad and township map of Iowa. Geo. F. Cram. Chas. Shober lith. 8 0782

1869
CHICAGO AND VICINITY
Blanchard's map of Cook and DuPage Counties. 4 1033

General plan of Riverside / Olmsted, Vaux & Co. Chicago Lith. Co. CHS

Guide map of Chicago. Rufus Blanchard. Chas. Shober lith. In *Citizens' guide for the city of Chicago.* 4 0418-19

Kimbark's addition to Hyde Park / Geo. W. Waite, C. S. Waite. 4 1403

Map of Chicago. Rufus Blanchard. 4 0422

Map of Chicago and environs. Rufus Blanchard. 4 0411

Map of Chicago and environs / Frank Davenport. Rufus Blanchard. Pfeiffer & Co. printers. 4 0414

Map of South Chicago and environs / Rufus Blanchard. Chas. Shober lith. From Clarke, Layton & Co.'s *Real estate: great annual sale.* 4 0412

Map showing Hill's addition to South Chicago / C. W. Stewart, Robert L. Conroy. 4 2676

New map of Chicago. Ed. Mendel lith. In Richard Edwards, *Directory to the...city of Chicago.* 4 0415

This plat shows the relative position of Oak Park, Harlem, and Thatcher to the city limits. In James B. Runnion, *Out of town.* 4 2361

Wilmette Village. Louis Nelke & Co. lith. 4 2898

ILLINOIS

New sectional map of the state of Illinois. Geo. F. Cram. Pfeiffer Co. printers. 4 1626

Rockford, Illinois. Western Bank Note & Engraving Co. In Kauffman & Burch, *Rockford city directory.* 4 2584

Sectional map of Illinois / H. F. Walling. Rufus Blanchard. 4 1631

OTHER

Cabinet map of the western states and territories. Rufus Blanchard. 1 1716

Cram's township & railroad map of the north western states. 1 0906

Map of Iowa. Rufus Blanchard. In his *Handbook of Iowa.* 8 0747

Map of Kansas. Rufus Blanchard. 13 0626

Map of Minnesota. Rufus Blanchard 7 0788

Map of Nebraska. Rufus Blanchard. In J. A. Bent, *Hand-book of Nebraska.* 12 0393

Map of the lands of the Union Pacific Railroad Company / H. F. Greene. Ed. Mendel lith. Two maps, for 1869 and 1870, in its *Guide to the Union Pacific Railroad lands* (1870). 12 1085-86

Map showing the mill sites and water power belonging to the St. Anthony Falls Water Power Co. [Minn.]. Chicago Lith. Co. 7 1505

New Map of Calhoun County [*Mich.*]. Merchant's Lith. Co. In E. G. Rust, *Calhoun County business directory.* 5 0316

New rail road & township map of Missouri and Kansas. Geo. F. Cram. 9 0917

New sectional map of the state of Iowa. Geo. F. Cram. 8 0791

New sectional map of the state of Missouri. Geo. F. Cram. Pfeiffer & Co. lith. 9 0916, 0918

Topeka, Capl. city of Kansas / J. B. Whitaker. Ezra A. Cook & Co. 13 1480

1860s

City of Knoxville, Illinois / N. Sanburn. Ed. Mendel lith. 4 1972

Map of Cass Co., Ill. / S. H. Burhans & J. Van Vechten. Ed. Mendel lith. BL

Map of the city of LaCrosse, LaCrosse Co., Wisconsin / Travis & Whittlesey. Ed. Mendel lith. 6 0669

1870

CHICAGO AND VICINITY

Buckingham's second addition to Lake View / Baird & Bradley. 4 1988

[*Map of Chicago*]. In John S. Wright, *Chicago past, present, future.* 4 0428

Map of Chicago and environs. Rufus Blanchard. 4 0424-25

Map of Chicago showing the location of South Lynne. A. Vail & Son. 4 0427

Map of Cook and DuPage Counties, Ill. / J. Van Vechten. 4 1034

Map of Englewood, Ill. J. Van Vechten. 4 1186

Map of the city of Lake View. Baird & Bradley. 4 0885

Outline map of Chicago. Western Bank Note & Engraving Co. 4 0429

Provident Homestead Association lots for sale. Horton & Leonard printers. 4 2342

ILLINOIS

Atlas map of Knox County, Illinois / Andreas, Lyter & Co. Chicago Lith. Co. 4 1969

Map of DeWitt County, Illinois. S. H. Burhans & L. M. Snyder. 4 1097

Map of Illinois. Rufus Blanchard. 4 1632

Map of Logan County, Illinois / S. H. Burhans & L. M. Snyder. Ed. Mendel lith. CHS

Miniature map of Illinois showing proposed boundaries of new dioceses. Chas. Shober lith. 4 1642

New sectional map of the state of Illinois. Geo. F. Cram & Co. Pfeiffer Co. printers. 4 1637

[Three maps of Illinois by James P. Root, showing representative, senatorial, and congressional districts]. Western Engraving Co. 4 1644

OTHER

Cabinet map of the western states. Rufus Blanchard. 1 0911,915

The city of Jackson, Michigan. Baker sc. In *The Land Owner,* vol. 2, no. 6. 5 1435

Cram's township & railroad map of the north-western states. 1 0913

East Saginaw, Mich. Baker sc. In *The Land Owner,* vol. 2, no. 9. 5 3352

Lansing, Michigan / M. B. Ladd. Baker sc. In *The Land Owner,* vol. 2, no. 8. 5 1633

Map of Ashland, Saunders County, Nebraska. Ezra A. Cook & Co. lith. 12 0015

Map of Iowa County, Wisconsin / E. T. & W. J. Wrigglesworth. Chas. Shober lith. 6 0577

Map of Kalamazoo Co., Mich. / Robt. S. Innes. Ed. Mendel lith. BL

Map of Kansas. Rufus Blanchard. In J. A. Bent, *Handbook of Kansas.* 13 0633

Map of Kansas / Ado Hunnius. Ezra A. Cook photo lith. 13 0637

Map of Plattsmouth, county seat of Cass County, Nebraska / George T. Nealley. Chicago Lith. Co. 12 1101

Map of the city of Grand Rapids, Michigan / E. W. Muenscher. Merchant's Lith. Co. 5 1079

Map of the state of Michigan. Merchant's Lith. Co. In Dillenback & Leavitt, *History and directory of Kent County, Michigan.* 5 2089

Map of Wisconsin. Rufus Blanchard. 6 1839

New railroad & township map of Michigan. Henry S. Stebbins. 5 2095

New sectional map of the state of Iowa. Geo. F. Cram. 8 0798

Outline map of Iowa, showing location of lands belonging to the Iowa Rail Road Land Company. Chicago Lith. Co. In its *Choice Iowa farming lands.* 8 0805

Rail road, county, and township map of the western states. Geo. F. Cram & Co. G. Ismael eng. 1 0914

Sectional map of Kansas. Rufus Blanchard. Chas. Shober lith. 13 0634

The town and harbor of Frankfort [*Mich.*] / Geo. S. Frost & Bro. Baker sc. In *The Land Owner,* vol. 2, no. 7. 5 0967

1871

CHICAGO AND VICINITY

Adam Smith's subdivision. J. S. Thompson & Co. 4 0142

Blanchard's map of Cook and DuPage Counties. 4 1035

Guide map of Chicago. Rufus Blanchard. In *Citizen's guide for the city of Chicago.* 4 0430

Map of Chicago / J. Van Vechten. Chas. Shober lith. G. W. Terry, map colorer & mounter. 4 0432

Map of Chicago and environs. Rufus Blanchard. 4 0431

Map of the north half of the town of Hyde Park / R. W. Dobson. Geo. W. Waite & Co. J. W. Middleton lith. 4 0929

Map of the towns of Hyde Park, Lake, Calumet, and the east half of Worth / J. T. Foster. J. Van Vechten. Merchant's Lith. Co. 4 0914

Plan of Lake Park, Chicago. J. M. Wing & Co. Baker sc. Land Owner eng. 4 1301

Plat of Cornell, Cook County, Illinois. John M. Wing & Co. Baker sc. Land Owner eng. 4 0928

Plat of Homewood and its residence sites. In *The Land Owner,* vol. 3, no. 8. CHS

Plat of Millard & Decker's improvements on Ogden Ave. Baker sc. Land Owner eng. In *The Land Owner,* vol. 3, no. 2. 4 0975

ILLINOIS

Atlas map of Fulton County, Illinois / Andreas, Lyter & Co. Chicago Lith. Co. 4 1242

City of Jacksonville; Holland's map of Jacksonville, Ill. In *Holland's Jacksonville city directory.* 4 1910

Combination atlas map of Stephenson County, Illinois / Thompson & Everts. Chicago Lith. Co. 4 2712

Map of Moline and vicinity with coal fields. J. M. Wing & Co. Baker sc. Land Owner Eng. 4 2242

Map of the Island of Rock Island. J. M. Wing & Co. Baker sc. Land Owner Eng. 4 2556

Map showing relation of Moline to the north-western states. J. M. Wing & Co. Baker sc. Land Owner Eng. 4 2243

OTHER

Fort Dodge, Webster County, Iowa. John M. Wing & Co. Baker sc. Land Owner eng. 8 0494

Lincoln, the capital of Nebraska / J. P. Lantz & Co. Chas. Shober lith. 12 0297

Map of Adrian, Michigan / Willits & Bird. G. W. Terry, map colorer & mounter. 5 0012

Map of Flint & Pere Marquette Railway Cos. lands in Michigan. Chas. Shober lith. 5 2711

Map of Kent County, state of Michigan / Sheldon Leavitt. Merchant's Lith. Co. G.W. Terry, map colorer & mounter. 5 1525

Map of Nebraska / State Board of Immigration. Rufus Blanchard. 12 0404

Map of the central part of western Iowa. Chicago Lith. Co. In Chicago, Rock Island & Pacific Railway Co., *Description of six hundred acres of choice Iowa farm land.* 8 0816

Map of the Cherokee neutral lands, Kansas / Missouri River, Ft. Scott & Gulf R. R. Co. Chas. Shober lith. 13 0954

Map of the Union Pacific Railroad land in Nebraska. Chas. Shober lith. In its *Guide to the Union Pacific Railroad lands.* 12 1087

New railroad & township map of Kansas and Missouri. Geo. F. Cram. 13 0651

New railroad & township map of Nebraska. Geo. F. Cram. 12 0400

New sectional map of the head of Lake Superior. Ed. Mendel lith. 7 1198

Plattsmouth, county seat of Cass County, Nebraska / D. H. Wheeler & Co. Baker sc. In *The Land Owner,* vol. 3, no. 2. 12 1102

Sectional map of Kansas. Rufus Blanchard. 13 0646

Sectional map of Kansas / Geo. O. Willmarth. Rufus Blanchard. 13 0653

Sectional map of Minnesota. Geo. F. Cram & Co. 7 0813

Sectional map of Nebraska. Geo. F. Cram. 12 0401

Waterville, Marshall County, Kansas / Central Branch Union Pacific Railroad Co. J. M. Wing & Co. Baker sc. The Land Owner eng. 13 1539